Microbubbles

Microbubbles

Engineering Aspects and Industrial Applications

Snigdha Khuntia
Ahmedabad University
Assistant Professor, School of Engineering and
Applied Science, Ahmedabad, India

WILEY

Published by John Wiley & Sons, Inc., Hoboken, New Jersey.
Published simultaneously in Canada.

Library of Congress Cataloging-in-Publication Data

Names: Khuntia, Snigdha, author.
Title: Microbubbles : engineering aspects and industrial applications / Snigdha Khuntia.
Description: Hoboken, New Jersey : Wiley, [2025] | Includes index.
Identifiers: LCCN 2024034342 (print) | LCCN 2024034343 (ebook) | ISBN 9781394249381 (hardback) | ISBN 9781394249404 (adobe pdf) | ISBN 9781394249398 (epub)
Subjects: LCSH: Microbubbles.
Classification: LCC QC161 .K47 2025 (print) | LCC QC161 (ebook) | DDC 530.4/275–dc23/eng20241010
LC record available at https://lccn.loc.gov/2024034342
LC ebook record available at https://lccn.loc.gov/2024034343

Cover Design: Wiley
Cover Image: © onuma Inthapong/Getty Images

Set in 9.5/12.5pt STIXTwoText by Straive, Chennai, India

Contents

Preface

Bubbles are fascinating and mysterious. When they are tiny, they are even more interesting to work with. Ever since I came across the fascinating world of tiny bubbles, which are famous in the name of microbubbles and nanobubbles, I have been blown away with their capabilities in several applications. Humankind is working hard to make every process and every industry to meet the norms of sustainability goals, and searching for alternatives to save the planet earth from the spread of pollution. I am intrigued to know the capabilities of the bubbles to achieve these goals of sustainability.

There are two major application areas of micro and nanobubbles: drug delivery and medical applications, and environmental applications. Medical and drug delivery applications of microbubbles are very different from the environmental applications. While drug delivery systems mostly work with loading and carrying the drugs between multiple media, their generation methods are also very different than environmental applications. For water treatment, flotation, and biofuel production, the major concern is to meet the scale of production. As a chemical engineer, I always look forward for a system that can meet the supply and demand of industrial processes, at the same time achieving minimal waste generation.

The microbubbles drive me to connect all the basic fundamentals of fluid flow operation, mass transfer operation, along with the chemical reaction engineering, which are the backbones of any chemical engineer. In the era of artificial intelligence and machine learning, there are several processes that still need attention to enhance their efficiency by doing research on the fundamental principles. Ever since the microbubbles have been discovered, numerous works have been published on their stability in the therapeutic and drug delivery applications. Microbubble-based systems have more potential than the conventional processes used in any multiphase reaction or non-reaction-based systems. To harness this potential, one needs to understand the physics and hydrodynamics of these amazing tiny bubbles, which are also very useful in the reaction medium. Most of the chemical engineers deal with several kinds of reactors in which bubble column reactors are the most common. Till date, many industrial processes are using the conventional macrobubbles or also known as millibubbles, which could be easily replaced with microbubble systems and achieve greater efficiency in their processes.

Earlier, the generation of microbubbles was difficult and the machines or generators were not energy efficient or easily accessible. Many engineers and entrepreneurs have now established many companies that now design and fabricate custom-made bubble generators with variable size and capacity. I believe more work and attention must be given to this amazing technology, which has the potential to reduce waste produced in the process industries and several chemical manufacturing industries. The major challenge faced by the water treatment industry is the amount of waste produced through several steps, where microbubble-based systems can be easily implemented and the waste production can be reduced up to a major extent.

When a student or an early researcher wants to explore more about the microbubbles in the application of water treatment and related fields, it is very hard to find a book on this topic. A beginner might find it difficult to start with research articles, which mostly adhere to a problem statement in a narrow research field. For graduate students, this problem is very evident where they find it challenging to relate a research topic with the undergraduate courses. This book will try to fill the gap of fundamental principles of microbubbles and their application in the industrial processes, where the students and researchers could find it interesting to work more with microbubbles. Although all the recent works and developments of the microbubbles in several applications cannot be covered in a single book, the basic principles, properties, and applications in the pilot-scale and large-scale operations have been covered.

I sincerely want to thank my PhD supervisors Professor Pallab Ghosh and Professor Subrata Kumar Majumder for introducing me to microbubbles in the application of water treatment, from which I primarily got the encouragement to research more on microbubbles. There are several pioneers in the field of microbubbles who always inspire me to work more on this topic, and the author thanks all of them. The author wishes to thank Ahmedabad University for providing support and facility to carry out the research and generous support in writing this book. Many thanks to my research team and students who helped me to collect several research articles and also during the writing process. I also thank my coauthors Professor Manish Kumar Sinha and Professor Abhishek Yadav who were willing to contribute two chapters, which were very essential to complete the scope of this book.

Despite all our efforts, there might be several errors and misleading sentences that might be confusing for the readers. Therefore, I urge all the readers to contact me without hesitation to add more value to this book.

26 November 2024

Snigdha Khuntia
Ahmedabad University
Assistant Professor
School of Engineering and
Applied Science
Ahmedabad, India

1 Overviews

1.1 Introduction

The pervasive threat of toxic chemicals in our society demands immediate attention. Despite the complexity of eliminating these hazardous substances, including the emergence of secondary compounds and economic constraints, safeguarding our environment remains paramount. With a particular emphasis on water quality as a linchpin of this endeavor (Hiroshi, 2006), the urgency to address this issue has never been more evident. As global environmental challenges escalate, a plethora of initiatives has arisen to bolster water purification efforts. Notably, microbubbles (MBs) have garnered increasing recognition within the sphere of water treatment technologies, signaling a promising avenue forward (Agarwal et al., 2010; Burns et al., 1997; Nakano et al., 2005; Wen et al., 2011).

MBs, tiny spherical bubbles filled with gas, exhibit remarkable individuality, resist aggregation, and typically range in size from 1 to 100 μm. Recent years have witnessed a notable upsurge in MB research, particularly in their application within various domains. These minute gas bubbles, when submerged in water, often measure less than 50 μm in diameter and are primarily composed of oxygen or air. Notably, MBs demonstrate an exceptional ability to remain suspended in water for extended periods. Gradually, the gas within them dissolves into the surrounding water, causing the bubbles to dissipate. MBs filled with air, oxygen, or ozone are extensively utilized across a diverse array of applications in the area of water treatment. Beyond that, MBs have been applied in numerous fields. For instance, in medical therapeutics, they serve as invaluable tools for organ imaging and as carriers for drugs or genes (Dicker et al., 2011; Kurup and Naik, 2010; Lindner, 2004). Moreover, MBs exhibit remarkable antibacterial properties, proving effective in both aerobic and anaerobic environments

Microbubbles: Engineering Aspects and Industrial Applications, First Edition. Snigdha Khuntia.

(Himuro et al., 2009). Additionally, various surfactant-stabilized MBs play a pivotal role in adsorbing proteins from water (Jauregi and Varley, 1998).

MBs have undergone functionalization with a variety of materials, spanning from surfactants and nanoparticles to pharmaceuticals and bioactive compounds. Their diminutive size endows MBs with exceptional efficiency in numerous industrial separation applications, notably in removing volatile contaminants and particulate matter (PM) from aqueous phases (Ahmed and Jameson, 1985). Both petroleum-based and biological surfactants have been employed to stabilize MBs, contributing to their versatility (Kukizaki and Baba, 2008; Xu et al., 2011). Flotation processes utilizing MBs hold particular significance in the separation of low-density PMs from water (Terasaka and Shinpo, 2007). In such processes, MBs serve as the carriers of PMs, lifting them from the column's base. The selection of a specific flotation method, whether it involves dissolved air flotation (DAF), dispersed air flotation, or electro-flotation, hinges on the technique utilized for MB generation (Liu et al., 2010).

MBs exhibit versatility across a multitude of fields, showcasing their wide-ranging applications. They have been integral technologies in cleaning processes (Akuzawa et al., 2010), extraction of oil from both soil and water (Gotoh et al., 2006; Xiaohui et al., 2011), and soil washing procedures (Roy et al., 1992). Moreover, MBs have played pivotal roles in fermentation practices (Ago et al., 2005; Xu et al., 2011), marine fish farming (Tsutsumi, 2010), and horticulture (Park and Kurata, 2009). In the realm of food technology, they have been utilized for various purposes (Shen et al., 2008; Soli et al., 2010). Furthermore, MBs have been employed in acid gas absorption, such as CO_2 using alkali substances (Akimov et al., 2011), and in the absorption of flue gases (Zeng et al., 2020), among numerous other applications summarized by Li (2006).

MBs exhibit an exceptional capacity to bolster the growth rate of marine organisms like oysters and scallops, facilitating efficient air transfer between phases through a substantial gas–liquid interfacial area. The adoption of MBs in water treatment has garnered substantial recognition across diverse industrial sectors, owing to their excellent efficacy as compared to the conventional methods (Jyoti and Pandit, 2001). In particular, ozone MBs have emerged as highly potent for oxidation, disinfection (Khuntia et al., 2012, 2013; Sumikura et al., 2007), decolorization of dyes and intermediates, and deodorization of water (Tasaki et al., 2009). Ozone, renowned for its robust oxidizing properties, proves invaluable in sterilization, color, and odor removal, as well as the degradation of organic substances in water purification and sewage treatment contexts (Camel and Bermond, 1998).

1.2 Generation Techniques of Microbubbles

The method chosen for MB generation plays a pivotal role in determining their properties and, consequently, their effectiveness in wastewater treatment. Achieving the production of fine MBs with a high density is essential for optimal wastewater treatment outcomes. Three primary types of microbubble generators (MBGs) find common use across various applications, and the types are gas–water circulation, pressurization–decompression, and no-liquid flow.

In gas–water circulation type, the required gas is introduced into a water vortex, where the turbulent flow transforms macrogas bubbles into MBs. Pressurization–decompression generators dissolve an ample amount of gas in water at moderate pressure. This process creates a supersaturated solution of gas in water. The gas escapes upon decompression from the water phase, generating a significant number of MBs. The third type, the no-liquid flow MBG, keeps the liquid static while only the gas is forced to mix, resulting in MB formation. Each of these generation methods is vital in customizing MB characteristics to fulfill the specific demands of wastewater treatment processes.

1.2.1 Types of Microbubble Generators

1.2.1.1 Gas–Water Circulation-Type Microbubble Generators

Rotational Liquid Flow Microbubble Generator (Spiral Flow Type) Ohnari (2000, 2002) led the pioneering development of the MBG, particularly the M2-LM type, engineered by Nanoplanet Research Institute Corporation. This cylindrical generator, measuring 50 mm in diameter and 100 mm in height, functions by introducing pressurized water from the bottom through a pump, initiating a rotational liquid flow within the cylinder. In agreement with Bernoulli's theorem, the center of the cylinder experiences reduced pressure due to the gyratory liquid flow, prompting the intake of air. This process generates an air-swirling flow, resulting in the production of MBs near the cylinder outlet. The flow rate of liquid is approximately 12 L/min, and the rotational speed of the two-phase flow in the cylinder ranges from 300 to 600 rounds/s. The ratio of gas–liquid flow rate is maintained between 1/7 and 1/15.

The practical implementation of the M2-LM-type generator has been observed in the aquaculture sector of oysters in Hiroshima, where it produces MBs of 10 to 50 μm size. However, it is to be noted that the density of MBs generated by the M2-LM method is relatively lower compared to that achieved by the DAF method. In an alternative setup, a centrifugal pump by Nikuni Corporation Limited was combined with the rotating-flow MBG M2-M from Nanoplanet Research Institute Corporation. This system simultaneously draws in water and air using the pump, with air dissolving into the water before reaching the generator. The pressurized water then experiences decompression through the MBG, which operates at a high rotational velocity. A similar but new MBG that works on the similar principle of rotational liquid flow was invented by Sadatomi et al. (2005), illustrated in Figure 1.1. The schematic of the MBG is shown in Figure 1.2, where the pressurized water enters the pipe having a spherical body situated in the core. The MBG works on the principle of Bernoulli's Theorem with conservation of energy and mass.

Venturi Type The venturi-type MBG operates with a liquid stream mostly water, entering the inlet of a venturi tube containing millibubbles. As

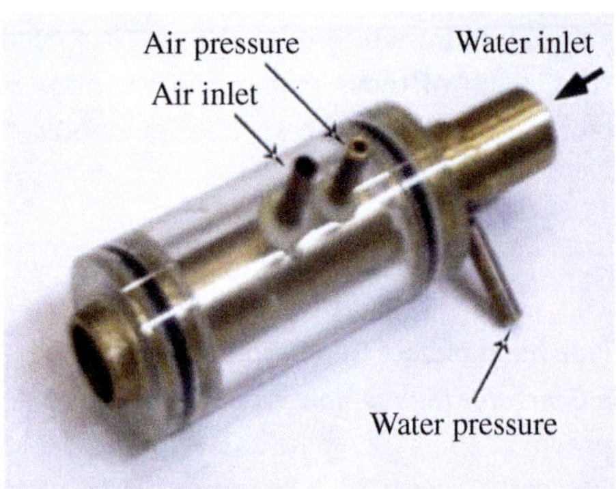

Figure 1.1 A rotational flow-type MB generator. Source: Reproduced with permission from Sadatomi et al. (2005)/ELSEVIER.

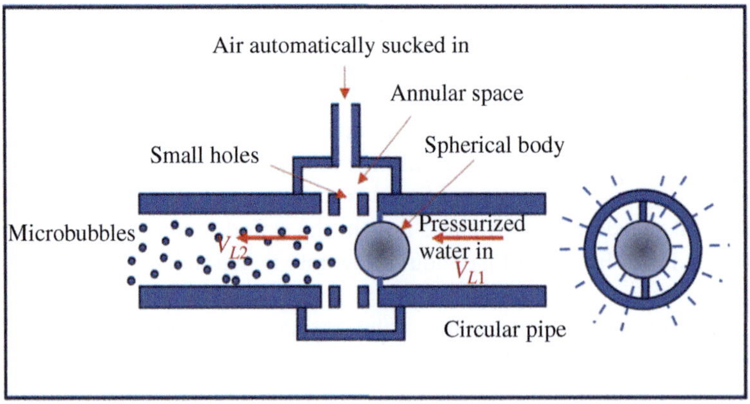

Figure 1.2 Schematic of MB generator. Source: Sadatomi et al. (2005)/with permission of ELSEVIER.

the two-phase flow propagates through the throat, rapid pressure changes occur, leading to the formation of MBs through the reduction of millibubbles and/or cavitation. In certain cases, to prevent bubble coalescence, 50 mg/L of 3-pentanol is introduced, resulting in the predominant production of 100 μm MBs. A schematic of spiral and venturi-type MBGs has been shown in Figure 1.3.

Static Mixer Type The static mixer-type MBG has been developed by Fluid Engineering Institute Corp., which creates helical multiphase flow using guide vanes and a special mushroom-type projection or current cutter. This transforms it into MBs ranging in size from 5 to 50 μm, achieving a maximum production rate of 1500 L/min.

Ejector Type The ejector-type nozzle-based MBG operates by expelling liquid at high velocity, functioning as an aspirator, where gas is absorbed due to opposing pressure at the outlet. This process initiates turbulent mixing and shearing mechanisms, ultimately resulting in the production of MBs. The average size of these MBs is approximately 424 μm, however, a little bigger than the conventional MBs.

Dissolved Air Flotation Type DAF MBG operates by dissolving pressurized air, typically at around 3–4 atm, into water. This air-infused water is then

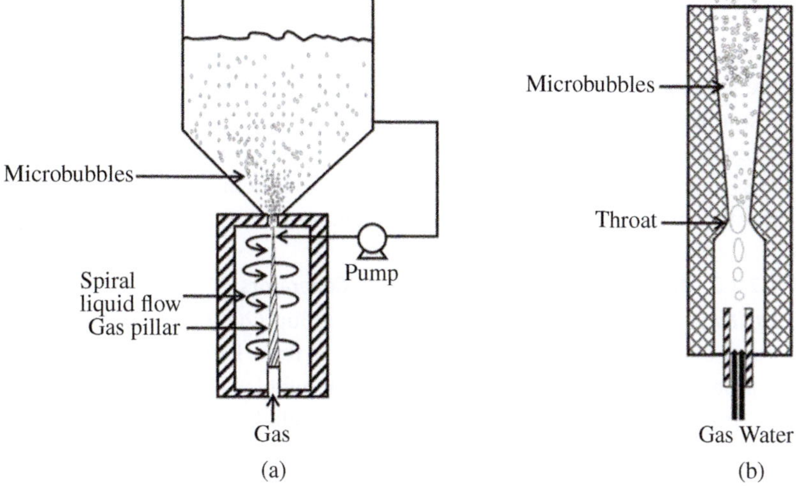

Figure 1.3 (a) Spiral type and (b) venturi-type MB generator. Source: Terasaka et al. (2011) / with permission of ELSEVIER.

discharged through the nozzle, generating MBs due to the supersaturation of air in the liquid. This process yields a high concentration of MBs, imparting a milky appearance to the liquid. The distribution of MB sizes typically exhibits two peaks, reflecting the complex interplay of factors involved in the generation process.

1.2.1.2 Pressurization–Decompression Type of Microbubble Generators

Gas dissolves by pressurizing into water within a tank to make the gas–liquid mixture. Upon flashing the supersaturated liquid through a reducing valve, MBs are generated, having size and quantity dependent on the decompression process. In a comparative study conducted by Terasaka et al. (2011), the performance of four MBGs was assessed, assessing gas holdup and gas mass transfer into the liquid phase. The spiral-liquid flow and the pressurization–decompression types of MBGs exhibited the highest gas holdup, with the former demonstrating a notable oxygen mass transfer coefficient. MBGs prove significantly more effective in transferring any gas into the liquid (mostly water) phase compared to commercial gas distributors such as perforated plates or constant-flow nozzles. However, the power demand for MBG is higher due to the additional pump attachment to create the pressure. Conventional gas distributors do not require this.

Ikeura et al. (2011) highlighted the effectiveness of the pressurization–decompression MBG in decomposing fenitrothion, a pesticide, using ozone. They attributed this superior performance to the tiny size of MBs generated by this generator variant, which facilitates enhanced contact of pesticides with water followed by the removal of excess pesticides in order to clean the vegetables. Variations in fluid flow patterns among different MBGs can result in differences in MB size, size distribution breadth, and MB shape. Moreover, some researchers have observed that the electrical properties of MBs may change depending on their generation method, which can significantly impact wastewater treatment outcomes.

1.2.1.3 No Liquid Flow-Type Microbubble Generators

Capillary Type Kukizaki and his research group (2009) have developed a novel sintered porous glass MBG fabricated from loamy soil, which is a capillary-type MBG. Along with MBs, nanobubbles (NBs) of diameter of 720 nm were generated from the sintered membrane, where the contactor has an average pore of size 84 nm. Various surface-active agents are often introduced to mitigate bubble coalescence.

Rotating Type In the rotating-type MBG, a high-speed rotation of a sintered material rotor within a surrounding stator automatically absorbs gas. Onoe et al. (2002) conducted reaction crystallization, yielding an average of 50 μm MBs. Notably, this generator eliminates the need for a liquid pump. Some of the available MBGs in this category are Fuki Works Corporation, Nomura Electronics Corporation, and Royal Electric Corporation, etc., which specialize in fabricating and selling rotating-type generators, mostly sold for the purification of lakes and other water bodies.

Ultrasound Type Yasuda et al. (2019) employed ultrasound irradiation on ultrapure water to create NBs of 90–100 nm in diameter. Achieving a high NB density was possible with the help of ultrasonication and irradiation time at a reduced ultrasonic frequency. In a study by Xu et al. (2008), common surfactants such as sodium dodecyl sulfate (SDS) and L-150 A (a mixture of ethanol 10% v/v, sucrose laurate ester 38% v/v, and water 52% v/v) were utilized for MB generation through sonication. Their findings revealed that using 1% SDS resulted in larger bubbles around 46 μm bubble. Additionally, the bubble density was 8.3×10^6/mL with SDS, while it increased to 2.5×10^7/mL with L-150 A. The impact of gas concentration on NB generation was also studied by Lee et al. (2020) through ultrasonic irradiation. They employed a 20 kHz ultrasonic horn booster, and they were subjected to ultrasonic irradiation for 10 minutes in three types of deionized water, such as undersaturated, saturated, and supersaturated. Their findings indicate that as the gas dissolution increased, the surface tension of water decreased, resulting in a reduced NB concentration.

Vapor Condensation Type In the vapor condensation-type MBG, nitrogen vapors and steam are mixed and blown into water from a nozzle. The steam condenses, yielding MBs, which is non-condensable nitrogen. The mean diameter of MBs ranges from 20 to 40 μm. The MB size is influenced by gas composition, diameter of the nozzle, and flow rate of mixed vapor.

Electrolysis Type MBs of O_2 and N_2 can be generated through electrolysis owing to redox reactions at the anode and cathode, respectively. Lucero et al. (2017) conducted studies to assess the impact of salt, current density, and distance between the electrodes on the rate of generation rate of MBs. The most dominating parameter was the current density that directly regulated the generation rate of MBs, with increasing current density leading to a more rapid bubble generation. The required energy for the electrolysis process is reduced by incorporating NaCl and reducing the electrode distance. In a study by Ulatowski et al. (2021), the effect of salt and duration of electrolysis on the mean MB diameter of N_2 and O_2 by electrolysis was investigated. Results indicated that the MB size increased with higher salt concentration, whereas only the hydrogen MBs were affected by the extension of electrolysis time and decreased in size with increased electrolysis time. Throughout the electrolysis process, the mean MB diameter initially decreased and attained an equilibrium size beyond an hour (Chandran et al., 2015).

1.2.2 Scale of Microbubble Generators

A variety of MBGs is now globally available, catering to diverse applications. In sectors such as wastewater treatment, fish culture, and related processes, there is a demand for plant-scale MBGs capable of meeting volumetric flow rate expectations while maintaining bubble diameter. Ozone is commonly used as the gas medium in MBs for many wastewater treatment processes, necessitating the MBG construction material to be corrosion-proof.

For laboratory-based studies, small pressurization–decompression and gas–water circulation types of MBG are prevalent. These generators can produce high-density MBs, with volumetric flow rates ranging from 5×10^{-4} m^3/s onwards. Nikuni, based in Japan, stands out as a leading manufacturer of various MBGs, supplying units worldwide. Their products find applications in laboratories, spas, and DAFs. The KTM model, specifically designed for DAF, boasts a water flow rate ranging from 22 L/min to 967 L/min. Another Japanese company, Hinode Sangyo Corporation Ltd., is also known for manufacturing pressurized MB systems. The company aims to offer a contemporary MB-based solution for wastewater treatment, with a focus on avoiding sludge formation in conventional processes.

Distinguished by its departure from the conventional MB generation methods, Aklova Microgas technology relies on direct MB induction instead of the principle of gas dissolution. This innovation results in energy-efficient systems that function with fewer components. Consequently, Aklova MicroGas^{TM}-powered plants boast easy installation, operation, and maintenance, featuring the smallest possible footprint. They claim to be the most convenient and highly energy-efficient MBG on the market for the DAF and gas transfer applications for O_3, O_2, and CO_2 gases. MicroGas^{TM} offers a gas flow rate range from 1–20 NL/min per unit for Series S, dedicated to laboratory and pilot-scale applications. Meanwhile, Series XL is designed for plant-scale applications with a gas flow rate requirement of 10–75 NL/min.

Among the widely encountered MBGs is the DAF type, readily available on e-commerce websites. Shanghai Edon Mechanical & Electrical

Equipment Co., Ltd. specializes in manufacturing the coupling-type micro-nanobubble (MNB) generator, catering to a diverse range of flow rates and applications. For applications often found in laboratory studies, YLEC Consultants have patented the unique MB-generating device named Carmin. The Carmin generator initiates gas MBs production by requiring a high-water inlet pressure of water flow for creating a cavitation pocket. Subsequently, gas flow in controlled fashion is injected into the vapor cavity, breaking up into an MB cloud of very uniform size. The typical water supply flow rate ranges from 1–26 L/min, and the injected air flow rate varies from 5–180 mL/min. One additional Japan-based company, OK Engineering Co., Ltd., specializes in providing nozzle-type MBGs designed for use in seawater without the risk of clogging. These nozzles can generate both MBs and NBs at a discharge rate ranging from 6 L/min to 25 L/min.

In the realms of cutting-edge technological innovations, MB technology has found widespread application in highly developed nations such as Japan, Korea, and China. Across the expanse of Japan, MB-generating systems have been made available through local manufacturers, including esteemed entities such as Aura Tech Company Limited, Shigen Kaihatsu Company Limited, and Nanoplanet Company Limited (Terasaka et al., 2011). While these developed MBs showcase remarkable technological prowess worldwide, their adoption faces hurdles beyond technological prowess. In contrast to their prevalence in advanced nations, the global dissemination of MB technology encounters challenges. Despite its technological advancements, it remains a relatively costly venture, fraught with logistical intricacies in transportation to developing nations. Its direct implementation proves intricate and, in some cases, deemed unfeasible, further complicated by the scarcity of spare parts. Indonesia, a nation known for its vibrant tapestry of islands, presents an intriguing case study. The application of MB technology for water treatment remains nascent, with its widespread adoption yet to transpire. Within the Indonesian landscape, MB technology has primarily found its niche in specific domains, notably in fishery industries and industrial labs of small scale (Deendarlianto et al., 2015; Majid et al., 2016). Remarkably, Indonesia lacks a common sparger manufacturing industry, a notable gap attributed to the absence of comprehensive research initiatives within the nation's borders. The relatively high operational costs of MB systems, stemming from the amalgamation of water pumps and air compressors (Terasaka et al., 2011), have deterred widespread implementation. However, the indisputable efficiency of MB technology, transcending its high energy consumption and cost of operation, beckons toward a future, where its transformative potential can be harnessed on a grand scale (Fan et al., 2018).

1.3 Physical Properties of Microbubbles

The fundamental distinction among macrobubbles, MBs, and NBs primarily hinges on their respective sizes. Macrobubbles typically exceed a diameter of 100 μm, while NBs are defined as those measuring less than 10 μm. Although there is no universally standardized size criterion, this size range is widely accepted worldwide. Over time, MBs gradually get smaller in size and eventually collapse due to prolonged stagnation. This process results in dissolution of the internal gases into the bulk water. In contrast, NBs have the remarkable ability to persist for extended periods, often months, without rupturing (Takahashi, 2009). Research has unveiled that the interface of NBs is marked by their robust hydrogen bonds, very similar to ice and gas hydrates. This unique property leads to decrease in diffusivity of NBs, contributing to their capacity to maintain a stable kinetic balance even in the presence of high internal pressure. For a visual representation, Figure 1.4 provides a clear schematic depicting the difference between macrobubbles, MBs, and NBs.

1.3.1 Physics of Microbubbles

Understanding the disparities between MBs and NBs based on their physical attributes and stability is of utmost importance. The MBs, while

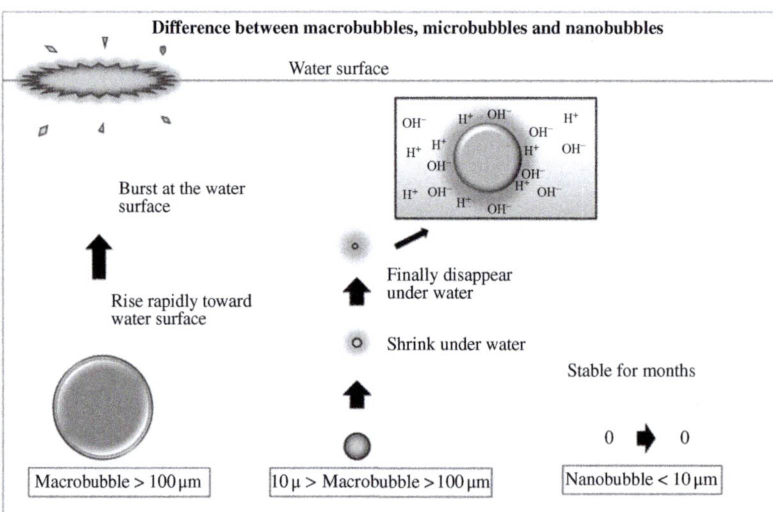

Figure 1.4 A schematic difference projecting the size difference of various types of bubbles. Source: Takahashi et al. (2007); reused with permission from the copyright holder, ACS Publishing.

dispersed in the water, give a milky appearance and then ascend to the surface slowly. Various techniques, including atomic force microscopy (AFM), have demonstrated the presence of NBs at the interface made of liquid–solid phases, which revealed that NBs at the interface take on the appearance of spherical caps with a height of 10 nm and a diameter of approximately 100 nm (Agrawal et al., 2005; Switkes and Ruberti, 2004; Zhang et al., 2006a). It has also been observed that the NBs can be fused together to create a larger bubble using the tip of AFM. Initially, it was believed that high surface tension of NBs was the reason behind this and assumed that the gas would quickly dissipate from them within microseconds of formation (Matsumoto and Tanaka, 2008). However, under the right conditions, NBs can be freely formed and maintain their stability for extended periods. The longevity of NBs is attributed to the lower interfacial curvature compared to initially anticipated, owing to a higher contact angle (Zhang et al., 2006b). Moreover, the formation of NBs in water containing small organic molecules, such as alpha-cyclodextrin, ethanol, tetrahydrofuran, and urea, has also been documented (Jin et al., 2007).

Bubble rising velocity and surface charge are the pivotal factors for comprehending the characteristics of both MBs and NBs. MBs were observed to migrate toward oppositely charged electrodes while observed in an electrophoresis cell. The surface charge, commonly known as zeta (ζ) potential, of each MB can be ascertained from its movement speed, which is directly linked to the magnitude of the ζ potential (Graciaa et al., 2000; Takahashi, 2005; Yoon and Yordan, 1986). In fact, the prolonged stagnation of MBs helps to determine the ζ potential readily in distilled water, which typically measures around −35 mV. It has been elucidated that MBs carry a negative charge over a broad range of pH values. It is worth noting that both OH^- and H^+ ions affect the charging mechanism of MBs at the gas–water interface (Takahashi, 2005). Furthermore, the ζ potential remains constant under similar water conditions and independent of the MB size, representing a consistent quantity of electrical charge per unit area. Nonetheless, an increase in the ζ potential is observed as MBs shrink during their collapse. A decrease in MB size below the surface of water leads to elevated internal pressure within the MBs, a pressure directly proportional to the diameter of bubble. The famous Young–Laplace equation describes the relationship between pressure (P), radius (r), and surface tension (σ) for a curved interface, commonly expressed as:

$$P_{\mathrm{G}} = P_{\mathrm{L}} + \frac{4\sigma}{d_{\mathrm{b}}} \tag{1.1}$$

This equation is used in various fields, including physics and engineering, to understand and calculate pressure differences across curved surfaces, such as bubbles and droplets.

In accordance with Henry's law, as gas pressure increases, the quantity of dissolved gas contained in a shrinking bubble also increases. Research has demonstrated that the region surrounding an MB undergoes a shift in its state within a pressure–temperature diagram, promoting hydrate nucleation (Sloan, 1998). This behavior stands as one of the distinctive attributes of MBs.

1.3.2 Microbubble Size and Measurement Techniques

MBs inherently assume a spherical shape. In the realm of three-dimensional objects at a constant volume, the surface area is minimized while the object attains a spherical form, a principle aptly described by the isoperimetric inequality (Osserman, 1978). Consequently, energy at the gas–liquid interface is at its minimum when a bubble adopts a spherical configuration. Millibubbles, frequently used in applications such as fermenters, gas–liquid or multiphase reactors, and DAF for mineral processing, typically feature diameters within the range of 2–5 mm. In contrast, MBs are characterized by diameters of 50 micrometers or less (Takahashi et al., 2007). Bubbles falling within the size range of less than 200 nm are designated as NBs, whereas bubbles ranging from 200 nm to 10 μm are referred to as "micro-nanobubbles" (MNBs) (Tsuge, 2010).

In an aqueous environment, MBs create a cloudy dispersion, whereas dispersions containing NBs exhibit transparency (Tsuge, 2010). According to Sebba (1988), a surfactant-stabilized MB consists of two concentric gas spheres, between which water layer is interposed, as illustrated in Figure 1.5. The inner gas core typically boasts a diameter of around 50 to 60 μm, and the internal water film has a thickness of approximately 1 μm. Surfactants contribute to the bubble's stability by introducing some interfacial forces that are repulsive in nature, such as the electrostatic double layer, hydration, and steric forces (follow Figure 1.5b). This stability enables the surfactant-stabilized MBs to be easy to transport without collapsing, allowing them to be generated in small vessels and subsequently pumped out into larger-volume reactors.

Numerous research articles have been published to study the effect of several factors on MB size. As reported by Wu et al. (2012), water temperature exerts a noticeable influence on MB size, primarily by affecting bubble coalescence. Elevated temperatures tend to promote the formation of larger bubbles. Alongside temperature, two other crucial factors influencing

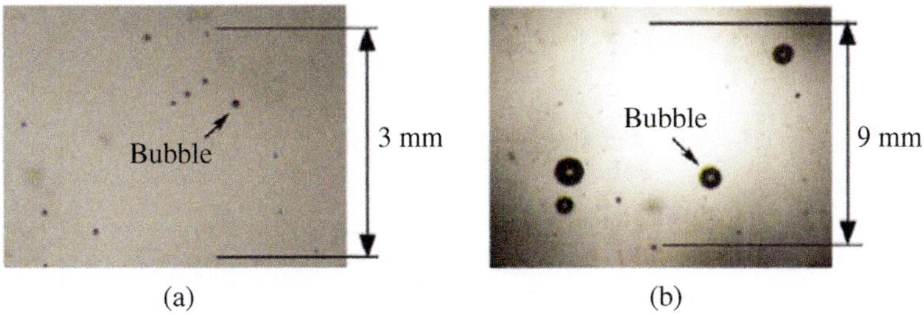

Figure 1.5 Images of Microbubbles (MBs) captured by, (a) bubbles below 0.1 mm and (b) bubbles above 0.1 mm. Source: Kawahara et al. (2009)/with permission of ELSEVIER.

bubble size are gas flow rate and pressure, as highlighted by Meegoda et al. (2018). Studies by Nazari et al. (2020) delved into the influence of gas flow rate on air MB size when utilizing a venturi tube, revealing that higher air flow rates, associated with increased dissolved gas, resulted in the generation of smaller bubbles. Temesgen et al. (2017) reached a similar consensus, observing that greater air density at higher pressure conditions led to the production of smaller bubbles. Moreover, Calgaroto et al. (2014) noted that the size of MBs exhibits some dependency on the magnitude of ζ potential, where at the isoelectric point the maximum bubble size is achieved. Table 1.1 provides a comprehensive overview of these pertinent factors, the associated operating conditions, and the methods employed to measure bubble size (Shen et al., 2008).

1.3.2.1 Photographic Method

Various methods have been employed for the measurement of bubble size and its size distribution. The simple photographic technique can be used to measure bubble diameter below 1 μm. The test apparatus setup is shown in Figure 1.6. To be precise, in images described in Kawahara et al. (2009), bubble clouds were captured at $h = 0.7$ m down from the surface of liquid using a digital Nikon camera (COOLPIX900.), which was equipped with a microscope lens (TelescoMicro Lens 8x20D, Nikon Co.) and a rear flashlight. For assessing bubbles lower than 0.1 mm within an area of 3 mm × 3 mm and bubbles larger than 0.1 mm within a 9 mm × 9 mm area, different magnifications were used, as illustrated in Figure 1.5. Each run entailed 10 for the larger area and 90 for the smaller area, ensuring that the total area examined remained consistent for both sizes. To measure the bubble size, an image processing system was employed using Cosmos 32, Library Corporation. This can detect a minimum bubble diameter of 10 μm. The major axis and minor axis of the bubbles in the images were measured;

Table 1.1 Various factors affecting the size of MBs.

Factors		Bubble size	Size measurement	Operating conditions	Reference
Solution temperature (°C)	20	200 nm	Mastersizer 2000 (Malvern Instruments)	NA	Ahmadi and Khodadadi (2013)
	25	230 nm			
	30	265 nm			
	35	310 nm			
	40	360 nm			
Gas pressure (kPa)	482	66.5 μm	High-speed, high-resolution flare camera + ImageJ software	The size measurements were performed with tap water and conducted under room temperature = 20 ± 2 °C	Gonzalez-Galvis and Narbaitz (2020)
Gas flow rate (L/min)	0.1	629 nm	Mastersizer 2000 (Malvern Instruments)	Liquid temperature = 32 °C, pH 8, gas pressure = 325 kPa	Nazari et al. (2020)
	0.18	482 nm			
	0.25	390 nm			
	0.33	300 nm			
	0.4	243 nm			
pH	2	130 nm	Zetasizer Nano ZS (Zen3600; Malvern Instruments)	Gas pressure = 455 kPa, bubbles were produced in 10^{-2} mol/L NaCl solution	Calgaroto et al. (2014)
	4	570 nm			
	6	715 nm			
	8	420 nm			
	10	300 nm			

Source: Shen et al. (2008)/with permission of ELSEVIER.

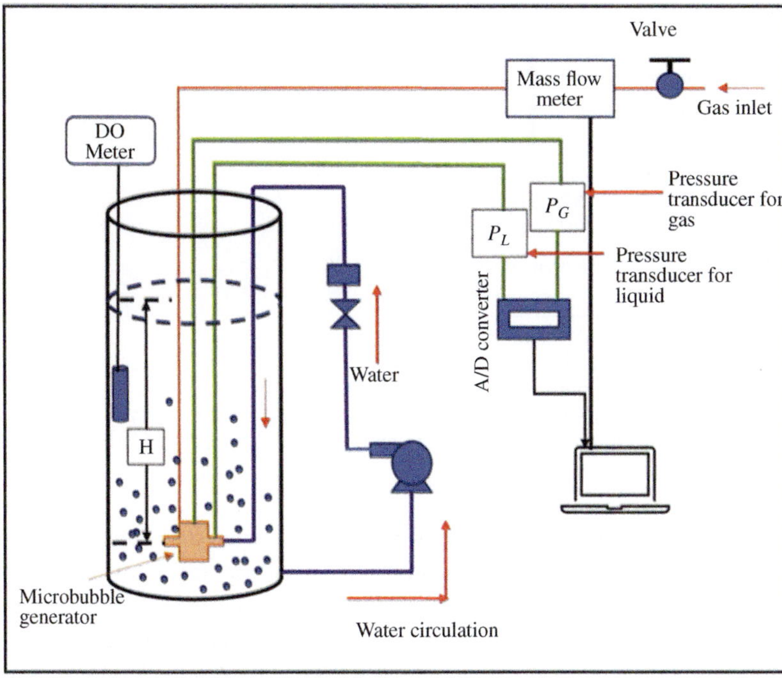

Figure 1.6 Test apparatus for generating MBs and capturing. Source: Kawahara et al. (2009)/with permission of ELSEVIER.

the average of these two axes was recorded as the bubble diameter, although most of them are spherical. Based on measurements of over 1000 bubbles, the Sauter mean diameter d_{BS} was calculated as follows:

$$d_{BS} = \frac{\sum n_i d_i^3}{\sum n_i d_i^2} \tag{1.2}$$

Li et al. (2016) utilized a similar method with another software, Image-Pro Plus, to assess the size distribution of both MBs and macrobubbles, ranging from 10 to 150 μm. ImageJTM is another widely adopted software for processing microscope images, as noted by Gonzalez-Galvis and Narbaitz (2020). However, generating visible images using this method can be cumbersome, given the limited field of view available under a microscope.

In a study conducted by Cavaro et al. (2011), MB generation was accomplished through the utilization of aeroflotation, a technique grounded in the concept of gas supersaturation, which triggers a phenomenon of the nucleation of bubbles. The histogram or the bubble size distribution of the generated bubble cloud was determined using an immersed optical system

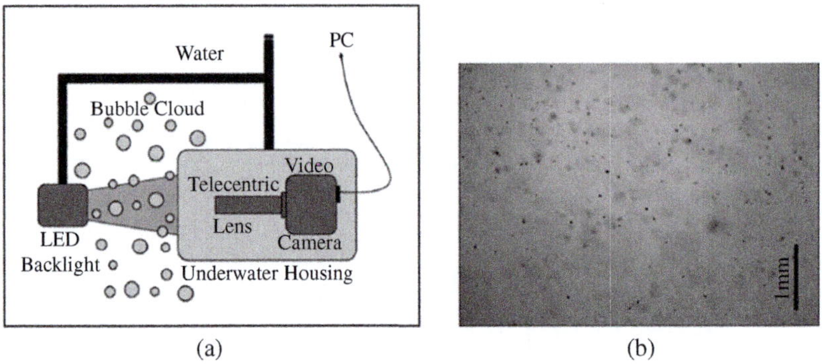

Figure 1.7 (a) The optical device to capture the photograph of MBs Source: Cavaro et al. (2011)/AIP Publishing/CC BY 4.0 and (b) the photograph of the MB cloud. Source: Reproduced with permission from Cavaro et al. (2011)/AIP Publishing LLC.

consisting of camera (5-megapixel) fitted with a telecentric lens, boasting a magnification of 1.5× and a focal length of 660 μm, in conjunction with an LED light source, as depicted in Figure 1.7a. This setup enabled the capture of images, such as the one illustrated in Figure 1.7b. To eliminate blurry bubbles, a dedicated "Image Processing" toolbox in MATLAB was employed. This process facilitated the reconstruction of the histogram for the observed bubbles. To validate the method, simultaneous measurements were conducted using a granulometric laser [Malvern, Orsay, France]. It is worth noting that a similar optical method employing a high-speed camera was used by Desai et al. (2019) (Photron FastCam S3), capable of capturing data at 250,000 frames per second with a resolution of 1024 × 1024 and equipped with a lens (Nikon Lens, 1.8 D). The region of interest encompassed 1.764 cm^3, taking into account the adjustment of different focal planes to ensure sufficient data capture.

Professor William Zimmerman and his team have reported studies comparing size distributions of MBs derived from multiple methods such as acoustic, optical visualization, and laser diffraction methods in their laboratory (Desai et al., 2023).

1.3.2.2 Dynamic Light Scattering (DLS)

Dynamic light scattering (DLS), as expounded by Berne and Pecora (2000), represents an advanced methodology for determining the diameter of MBs and NBs. This approach quantifies the Brownian motion to establish its correlation with bubble size. As bubble size increases, the Brownian motion perceptibly slows down. The rate of this motion is precisely characterized by the translational diffusion coefficient. To estimate bubble size from the translational diffusion coefficient, Stokes–Einstein equation is used (Ghosh, 2009a).

$$d_b = \frac{RT}{3\pi \mu_L N_a D_b} \tag{1.3}$$

Meegoda et al. (2018) have reported the calculation of d_b by using Zetasizer Nano ZS, an instrument from Malvern Instruments Ltd., based on DLS to analyze the size distribution of both MBs and NBs spanning a size range from 0.3 nm to 10 μm. On the other hand, da Cruz et al. (2016) employed another device from Malvern Instruments Ltd., Mastersizer 2000SM, which is a laser diffraction instrument, to determine the mean bubble size. Notably, DLS outperforms LD when it comes to measuring samples as small as MBs and NBs, as noted by Badger (2019).

The diameter determined through DLS corresponds to the particle's diffusion in liquid and is known as hydrodynamic diameter. This value represents the sphere diameter exhibiting the equal translational diffusion coefficient of the bubble. Notably, it depends on the bubble size as well as on factors such as surface structure, ionic concentration, and the types present in the medium. Various elements influence the rate of particle diffusion, including the medium ionic strength, which impacts the electrostatic double layer and its thickness, the structure of the surface, for instance, and the presence of long-chain molecules absorbed to the surface of the bubble entering the liquid. Additionally, fluctuations in the medium temperature can induce changes in viscosity, potentially leading to convection currents that affect the random motion of molecules. Therefore, maintaining a stable temperature is imperative in DLS experiments.

Laser light scattering represents an advanced approach to gauging the dimensions of MBs. A photomultiplier tube is employed to capture the scattered light. The connection between the bubble diameter and the scattered light intensity varies according to bubble size. For NBs, one can apply the Rayleigh scattering theory (when bubble diameter is less than the light's wavelength), where the scattered intensity is directly proportional to the sixth power of bubble diameter (Pelssers et al., 1990). Conversely, when dealing with larger bubbles, the Mie scattering theory is utilized to measure bubble diameter, where the bubble diameter is closer to the light's wavelength (Kerker, 1969). According to Glantschnig and Chen (1981), bubble diameter and the scattered light intensity exhibit a complex relationship, wherein for large MBs, it varies with the square of the MB diameter. Laser scattering technique proves invaluable for quantifying bubble size distribution within a range spanning from 0.02 μm to 2000 μm. Calibration of size and scattered intensity is conducted with the aid of standard particles like polystyrene latex particles. When laser light interacts with spherical bubbles, it produces a scattering pattern. Notably, bands of interference may also become visible on a plane distant from the focal point. The MB diameter is directly proportional to the number of interference bands (Tsuge, 2010). Consequently, this technique facilitates the measurement of MB diameter spanning from 10 μm to 1 mm.

1.3.2.3 Laser Diffraction

Low-angle laser light scattering (LALLS), or simply laser diffraction, was initially introduced in the mid-1970s, but it has seen significant advancements in instrumentation over the past couple of decades. LD is categorized as a nondestructive as well as a nonintrusive technique, relying on the principle that the angle of laser diffraction is inversely proportional to the size of the particles being analyzed. Its applicability covers a broad size range, typically spanning from 0.01 to 3500 μm, though it is worth noting that ISO13320-1 (1999) narrows it to 0.1–3000 μm. In the context of mineral samples, it is crucial to consider hydrodynamic limitations that impose restrictions on the upper size limit. These limitations primarily revolve around the challenge of maintaining bigger and heavier suspended particles without altering their size distribution.

A general laser diffraction system comprises a helium–neon light source wavelength fixed at 632.8 nm, along with appropriate detectors to capture the pattern of the light scattering. Signal processing and results output are handled by a personal computer (Xu, 2002). Laser diffraction allows for the rapid assessment of size distributions, typically yielding results within seconds, and complete analyses can be completed in less than a minute (Rawle, 2002). The results obtained through laser diffraction are expressed as volume-equivalent-spherical diameters (ESDs). Notably, when dealing with MBs, which have diameters below 100 μm and are often considered rigid spheres (Clift et al., 1978; Edzwald, 1995), there is no need for stereological conversions, as is common with most solid particles (Schneider et al., 2007). It is worth mentioning that laser diffraction techniques have been previously applied to measure the size of MBs in colloidal gas aphrons, a type of foam commonly encountered in separation processes (Bredwell and Worden, 1998; Couto et al., 2004). This prior experience motivated the use of laser diffraction in this work for assessing size distributions in the DAF process. Once again, it is important to note that laser diffraction-based size distribution assessments are remarkably quick, with complete analyses being accomplished in less than a minute.

To enable the comparison of results obtained through various technical approaches, d_b, or the mean diameter of MBs, was first estimated using fluid dynamic correlation, a method outlined by Couto et al. (2004). This estimation involves the use of Eq. (1.4), derived from the well-established Stokes' law that accounts for the influence of bubble density on the liquid side volume fraction (ϵ_L). This concept was first introduced by Richardson and Zaki (1997). Equation (1.4) is particularly suited for application when the Reynolds number (Re) is below 0.2, corresponding to the Stokes regime, and serves as a valuable tool in this context.

$$d_b = \left[\frac{18 V_0 \mu_L}{(\rho_L - \rho_G) g \epsilon_L^{4.65}} \right]^{\frac{1}{2}} \tag{1.4}$$

The variable ϵ_L was determined experimentally using the methodology detailed in Couto et al. (2004, 2009). The approach utilized here primarily involves measuring the rising velocities of the bubble front, denoted as V_i and V_f, within a graduated cylinder. This enables the determination of ϵ_L through the ratio V_i/V_f. It is crucial to highlight that this technique measures the average bubble diameter. This is made feasible by accurately measuring the ascending speed of the bubble front during the initial phase of the test, typically lasting around 10–20 seconds, while the interface is visible. Within this short timeframe, the bubbles rise closely to one another, making their diameters reasonably representative of the mean. However, as time progresses, larger bubbles rise more rapidly than the smaller ones, potentially introducing interference with the measurements.

1.3.2.4 Particle Image Velocimetry (PIV)

The particle image velocimetry (PIV) method employed for MB generation in the bubble column involved recording the process in a darkroom illuminated by a source of scattered light generated by a laser diode module, specifically the CPS635R type with a wavelength of 638.1 nm. The MB generation was captured in video format using a Sony RX 100 digital camera operating at up to 1000 fps high frame rate (Alam et al., 2020). The MBs characterization process using the PIV has several steps: capturing the video images showing the rising movement of the MBs, image acquisition followed by image extraction or preprocessing, establishing bubble displacement calibration from pixels to microns, conversion of displacement to measurement of particle velocity, and finally estimation of particle size followed by analysis of the size distribution of bubbles.

Spatial digital image correlation (SDIC) was employed to determine the MB rising velocity, a technique previously developed for characterizing the size and distribution of MBs. In this method, using a laser diode light source, the recording of the scattered light from the MBs is performed using a digital camera. Consequently, the accuracy of the method was validated through a repeatability test conducted five times. Utilizing the Δt between each Δs frame, the size of MBs can be calculated following the Hadamard–Rybczynski equation:

$$r = \sqrt{\frac{3}{2}\frac{\mu_L}{g}\frac{(2\mu_L + 3\mu_G)}{(\rho_L - \rho_G)(\mu_L + \mu_G)}\frac{\Delta s}{\Delta t}} \tag{1.5}$$

This approach enhances the precision of MB characterization and contributes to a comprehensive understanding of their dynamics in the given experimental setup.

1.3.2.5 Other Methods of Microbubble Size Measurement

Various other methods are also available for measuring bubble size and size distribution, including techniques such as the pore electrical resistance method, phase Doppler anemometry, and the acoustic bubble spectrometer. In the method reported by Tsuge (2010), which is pore electrical resistance method, a setup involves placing electrodes in an electrolyte solution divided with the help of a separator with small pores. By applying voltage to the electrodes, the electrical resistance is measured, primarily depending on the resistance occupied by micropores. As MBs pass through these pores, they cause an increase in electrical resistance, allowing for the determination of both the number of bubbles and their sizes. This method is effective for measuring bubble diameters within the range of 0.2–600 μm. However, it is worth noting that if the distribution of bubble diameters extends beyond this range, adjustments to the pore size may be necessary for accurate measurements. Hosokawa et al. (2009) introduced a novel approach for measuring MB diameter, which combines phase Doppler anemometry coupled with image processing technique, as described by Yu et al. (2007). This innovative method allowed for the evaluation of MBs' mean diameter and size distribution, with the results being derived from measurements obtained through both techniques. The authors have asserted that this approach offers accurate determinations of size distribution across a wide range of diameters.

In a different study, Rehman et al. (2015) employed an acoustic bubble spectrometer to assess the bubble size distribution, encompassing MBs from 80 to 450 μm. Acoustic spectrometry exhibits a superior capacity for detecting MBs and NBs in comparison to laser diffraction and optical techniques, according to the findings of Desai et al. (2019).

1.4 Characterization of Microbubbles

1.4.1 Size Distribution

Specific surface area of MBs and NBs is more than the macrobubbles, which can be determined using the following equations:

$$S = 4\pi r^2 \tag{1.6}$$

$$V = \frac{4\pi r^3}{3} \tag{1.7}$$

$$\frac{S}{V} = \frac{3}{r} \tag{1.8}$$

MBs, whether generated ultrasonically or through specialized generators, typically exhibit a broad size distribution. The creation of monodispersed MBs, as highlighted by Zhang and Li (2010), poses a considerable challenge. Monodispersed MBs hold significant importance in medical applications. MB size distribution can assume either a unimodal or a bimodal form, with a representative size distribution (Feshitan et al., 2009; Tsuge et al., 2009). It is crucial to acknowledge that the technique employed for MB generation profoundly influences their size, as explained by Li (2006). In a study conducted by Tsuge et al. (2009) on single-pore nozzles and rotating-flow MBGs, they made a direct comparison of the sizes of bubbles they produce. With the rotating-flow MBG, the MBs undergo shearing and crushing from the rotating flow, resulting in much smaller MBs. Moreover, at higher outlet pressures from the pump, both generators produce smaller bubbles. This phenomenon is attributed to the pressure change during bubble formation, leading to greater pressure differentials that yield smaller bubbles.

The size distribution of MBs is typically represented as the cumulative distribution, in which the percentage of the total MBs volume below a specific size is drawn against the size of bubble. Specifically, d^{90}, d^{50}, and d^{10} correspond to the 90th, 50th, and 10th volume percentiles of the cumulative size distribution profile. Another factor, the dispersion coefficient of bubble size (δ), is formally defined as (Kukizaki and Goto 2006):

$$\delta = \frac{d^{90} - d^{10}}{d^{50}} \tag{1.9}$$

The z-average diameter of MBs is measured by DLS. This z-average diameter represents the mean diameter within the distribution, and it is derived from the cumulative analysis of the correlation function, which is dependent on the intensity of the scattered light. It is essential to note that this diameter is independent of mass or number average, as it is evaluated from the signal intensity. In the context of DLS, this value holds paramount significance and offers a high degree of stability. For quality control purposes, this *z*-average diameter is the preferred metric. It becomes directly comparable with measurements from other techniques when the bubble distribution is unimodal and the bubbles conform to spherical and monodispersed characteristics. The cumulative size distribution reported by Couto (2009) is shown in Figure 1.8.

The influence of surface-active impurities and electrolytes on MB size is significant. As reported by Walker et al. (2001), an increase in NaCl concentration up to 1 mol/L leads to a decrease in size of MBs. Additionally, observations reveal that the elasticity of the surface plays a pivotal role in the generation of finer bubbles in the presence of salt (Hofmeier et al., 1995). In cases where pure liquids with low surface elasticity are involved, the result is the formation of larger bubbles due to coalescence. Conversely, in aqueous solutions characterized by higher surface elasticity, bubble size was reduced owing to the reduction of coalescence.

1.4.2 Hydrodynamic Properties

The primary objective of measuring bubble size is to accurately predict bubble characteristics, including rise velocity, and establish correlations with heat and mass transfer phenomena. Table 1.2 outlines several methods

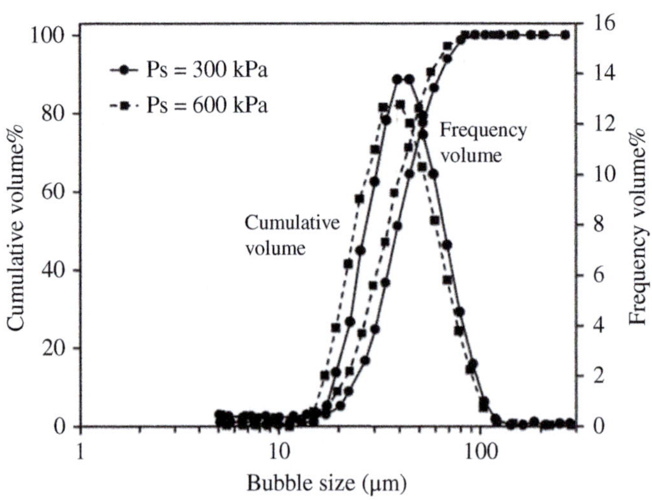

Figure 1.8 The experimental observations of MB size distribution with respect to various saturation pressures, measured using laser diffraction technique. Source: Couto et al. (2009)/with permission of ELSEVIER.

Table 1.2 Available literature on the techniques for MB size measurement and MB rise velocity.

Bubble size measurement				
Measurement technique	Experimental setup	Analysis	Size range (μm)	Reference
Light diffraction	Samples drawn into laser online particle counter	Conversion of diffraction to bubble size	15–85	Han et al. (2002)
Electrical resistivity	Samples drawn into batch-type particle counter	Conversion of resistivity to bubble size	13–96	
Digital camera and backlighting	Lab scale dissolved air flotation (DAF) unit with samples drawn into viewing chamber	Bubble size analyzer software LabVIEW (BASF)	60–131	Gulden et al. (2018)
Photography	Samples tapped into Perspex viewing cell (0.08 m, 0.08 m, and 0.015 m)	Image analysis software Image-Pro plus	10–150	Li et al. (2016)
Photography	Samples from DAF unit tapped off into cuvette	Automatic image analysis	10–300	de Rijk et al. (1994)
Gas velocity and gas holdup	Pilot flotation column	Drift flux analysis	350–1100	Filippov et al. (2000)
Acoustic spectrometry	Bubbles generated by diffusers in tank	Analysis of signals obtained by acoustic bubble spectrometer system using software	80–500	Rehman et al. (2015)

(*continued*)

Table 1.2 (Continued)

		Bubble size measurement		
Digital camera	Lab-scale flotation unit	Stochastic image analysis incorporating bubble clusters	200–2000	Kracht et al. (2013)
	Flotation cell	image analysis software	500–3000	Grau and Heiskanen (2002)
	McGill bubble size analyzer (sampling tube & tilted viewing chamber)	One-dimensional discrete Fourier analysis		Vinnett et al. (2018)
		Bubble rise velocity		
Measurement technique	Experimental setup	Analysis	Size (μm); rise velocity (mm s^{-1})	Reference
High-speed camera	Single bubbles transferred into a cuvette	Manual image analysis	10–120; 1–12	Parkinson et al. (2008)
Photodetector	Single bubbles generated in electrophoresis cell	Laser Doppler velocimetry	≈80; 4–5	Kelsall et al. (1996)
High-speed camera	Carried out batch flotation of bubble-particle flocs in jar tester	Particle image analyzer software	200–700; 9–15	Kwak et al. (2009)
High-speed camera	Single bubbles transferred into a viewing chamber	MATLAB image analysis	1300–2000; 200–500	Eskanlou et al. (2018)
High-speed camera	Downward flow chamber used to isolate a bubble and keep it stationary relative to camera with flow stopped periodically for velocity determination	Manual analysis using the known distance traveled	1000–5000; 100–350	Alves et al. (2005)

employed to characterize MB size distributions and rising velocities. The most critical steps in bubble characterization experiment involve (i) acquiring samples of bubble-filled liquid; (ii) measuring size and rise velocity of the bubble; and (iii) conducting the data analysis.

While imaging techniques have been widely utilized, alternative techniques are (i) light diffraction, (ii) drift flux analyses, (iii) porous plate, (iv) electro resistivity, and (iv) optical detectors. Data analysis methods range from manual measurements of individual bubble sizes to even more sophisticated automated approaches capable of determining the bubble sizes within clusters. Notably, rise velocity of MBs measurement frequently utilizes high-speed photography coupled with image analysis, as highlighted by Swart et al. (2020). Chapter 3 discusses the parameters measured to determine the hydrodynamic properties of MBs.

1.4.3 Electrical Properties

The electrical potential, commonly denoted as the zeta (ζ) potential, of suspended particles such as gas bubbles or oil droplets plays a pivotal role in revealing the interplay of attractive and repulsive forces among them. This zeta potential manifests at the interface between the particles and the surrounding liquid, substantially influencing the stability of colloidal dispersions. The value of the ζ potential serves as a key indicator of this stability. A higher absolute ζ potential value indicates a greater resistance to agglomeration, thus rendering the dispersion or solution more robust (Shangguan et al., 2018). In general, in distilled water, bubbles tend to carry a negative charge due to the hydroxyl ions (OH^-) adsorbed at the interface (Bui et al., 2019). Table 1.3 provides a summary of several research studies reporting the impact of several factors, including the type of gas, gas pressure, pH of the solution, and the introduction of surfactants, on ζ potential.

1.4.3.1 Effect of Gas Type

The ζ potential varies based on the gas utilized for bubble formation. Oxygen and nitrogen as injection gases result in notably high ζ potential for both MBs and NBs when compared to other injection gases such as air, carbon dioxide, and xenon, as reported by Ushikubo et al. (2010). In a similar vein, Meegoda et al. (2018) employed N_2, O_2, O_3, and air for NB generation in de-ionised (DI) water. Among these, the O_3-MBs exhibited the highest ζ potential, followed by O_2, air, and N_2. This variance in ζ potential is attributed to differences in the rate of gas diffusion and solubility, which, in turn, affect the adsorption of OH^- ions at the surface of the bubble.

Table 1.3 Factors affecting the ζ potential of MBs.

Factors		ζ potential (mV)	Reference
Gas type	Nitrogen	−19.9	Meegoda et al. (2018)
	Air	−21.3	
	Oxygen	−22.4	
	Ozone	−27.0	
Pressure (kPa)	138	−0.3	Ahmed et al. (2018)
	207	−3.8	
	275	−10.0	
	345	−15.2	
	414	−22.5	
pH	5	−18.2	Kim and Kwak (2017)
	7	−21.8	
	9	−28.1	
	11	−30.5	
Surfactant concentration (anionic) (mg/L)	50	−37.8	Jia et al. (2013)
	100	−42.0	
	150	−52.2	
	200	−64.1	

The ions at gas–water interface hold a pivotal role in preserving the stability of MBs, preventing neighboring MBs in the dispersion (Ghosh, 2009). Additionally, it is instrumental in facilitating the generation of hydroxyl radicals (·OH), which hold immense significance in the oxidation of mostly organic and some inorganic pollutants in water and wastewater (Bando et al., 2008; Takahashi et al., 2007). In distilled water, the air-MBs show consistently negative ζ potential. Takahashi (2005) reported a value of −35 mV, while Qu et al. (2009) found it to be approximately -57.9 ± 3.9 mV at moderately acidic pH = 5.8. Notably, the latter researchers introduced a minute quantity of salt (NaCl = $0.05\,mol/m^3$) in the distilled water. Generation of O_3-MBs using two MBGs was reported by Hasegawa et al. (2009), featuring slit angles of $\pi/3$ and $\pi/6$ radians. They observed only a little deviation in the values of ζ potential, specifically −43 mV and −38 mV, for two different slit angles.

1.4.3.2 Effect of Addition of Surfactants

Jia et al. (2013) delved into an exploration of how surfactant type and concentration affect the ζ potential of air bubbles within an aqueous solution. Their findings revealed that the introduction of a cationic surfactant had the effect of diminishing the –ve charge associated with the ζ potential, and in some instances, it even transformed the –ve charge into a positive one. Conversely, the inclusion of an anionic surfactant intensified the negativity of the ζ potential. The ζ potential, in this context, is determined by gauging the electrophoretic mobility of the MBs and subsequently using the Smoluchowski equation, as elucidated by Ghosh (2009).

$$\zeta = \frac{\mu_L E}{\epsilon \epsilon_0} \tag{1.10}$$

The shrinking of the MB helps in the generation of radicals, depending on the gas medium. The detailed mechanism and studies for the generation of radicals have been discussed in Chapter 4. The coated MBs were generated using amphiphilic surfactants by Shen et al. (2008), as shown in Figure 1.9. The MBs were 0.1–200 μm in size, which were generated by using four different types of surfactants, such as monoglyceride, sodium stearoyl-2-lactylate, diglyceride, and PEG 40 stearate. Figure 1.9 (a) shows the double-layer structured surfactant-coated MBs, whereas Figure 1.9 (b) shows a photograph of the same at different time scales.

1.4.3.3 Effect of pH of Water

The ζ potential exhibits a profound sensitivity to the pH of water. According to Takahashi (2005), the ζ potential of MBs in distilled water is negatively charged when the pH exceeds 4.5. At highly alkaline conditions with a pH greater than 10, the ζ potential surpasses –100 mV. In contrast, when the pH falls below 4.5, the ζ potential becomes positive. The charge that develops at the gas–water interface arises from the adsorption of both OH^- and H^+ ions. These counterions are drawn near the interface, forming a double layer (also known as electrostatic double layer). In alkaline solutions with a high pH, the OH^- ions tend to adsorb at the interface, leading to a negative charge on the MB. Conversely, in acidic solutions, there is an excess of H^+ ions over OH^- ions at the interface, causing a positive ζ potential. It is important to note that the presence of inorganic salts, such as NaCl and $MgCl_2$, can modulate the ζ potential, making it less negative, and the extent of this effect depends on the concentration of these salts (Takahashi, 2005).

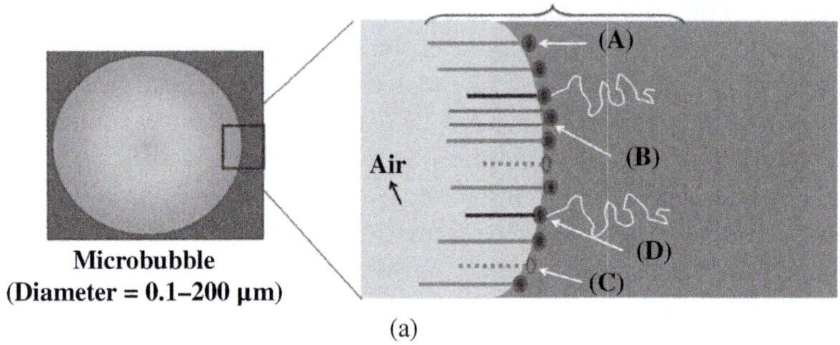

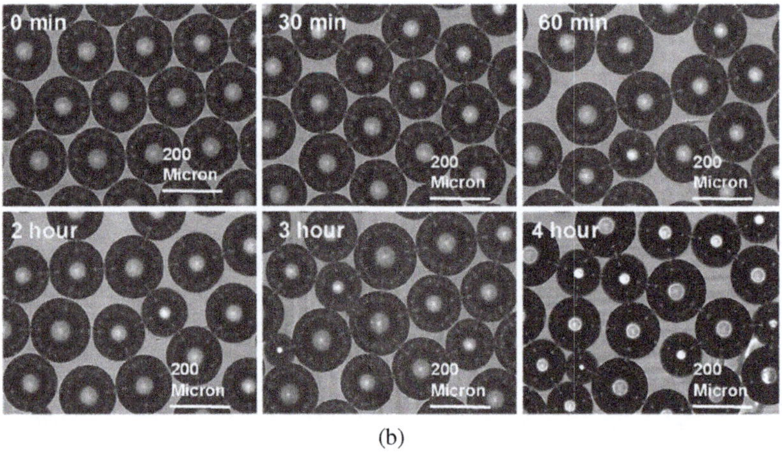

Figure 1.9 Surfactant-coated MBs: (a) schematic structure of MBs and (b) photograph of the MBs. Source: Reproduced with permission from Shen et al. (2008)/ELSEVIER.

The comprehensive summary of various factors affecting the ζ potential of MBs has been listed in Table 1.3.

1.4.4 Coalescence and Density Control

Generating high-density MBs with a diameter of 100 μm or less poses a considerable challenge, as they tend to merge upon contact. Research suggests that the Marangoni effect, induced by a spiked surfactant with molecules adsorbed onto the bubble surface, inhibits bubble coalescence. To enhance MB density, a minute quantity of a specific alcohol compound (10–1000 mg/L) was found to be crucial. In a study reported by Miyamoto et al. (2007), the researchers observed that this compound effectively prevented bubble merging. The presence of additive exhibited cloudiness, indicative of a high MB density, while without the additive the solution remained transparent due to the formation of larger bubbles, measuring several millimeters in diameter.

The functioning mechanism of the additive remains unclear, despite its proven efficacy. However, other authors have given the reason of reduced surface tension for this phenomenon. To elucidate this mystery, the authors meticulously observed bubble formation from micropores captured with a high-speed microscopic video camera. The absence of the additive makes the bubbles small in size in pure water, but their rapid merging and growth in size upon collision became evident. Conversely, in pure water addition of 0.01% alcohol, MBs exhibited a distinct behavior, reminiscent of balloons rebounding off each other upon collision. The authors conducted tests with various alcohol compounds featuring similar structural isomers, revealing considerable differences in their ability to generate MBs. A general hypothesis is that alcohol molecules adsorb onto the MBs surface, preventing their merger. Moreover, even a small chemical structure difference is proposed to exert a significant impact on MB generation. Essential additive characteristics encompass water solubility, minimal environmental impact, exemption from the purview of any other legislation, low volatility, and independence from temperature fluctuations. In the absence of the additive, bubbles had a diameter of approximately 4 mm. Upon adding a trace amount of the alcohol in pure water, the mean MB diameter changed to 75 μm, consequently increasing the surface area by 50 times or more.

The decline in size is correlated with an elevation in surfactant concentration, a relationship explicated by an empirical equation employed by Parmar and Majumder (2014):

$$d_\mathrm{b} = 52.05 \times 10^6 \exp(-0.03C_\mathrm{s}) \tag{1.11}$$

This equation encapsulates the dependence of bubble diameter on surfactant concentration. As the surfactant concentration increases, the liquid surface tension diminishes, leading to a reduction in bubble size. Confirming this trend, researchers such as Sardeing et al. (2006) and Takagi et al. (2008) have also noted that surface-active agents not only diminish surface tension but also yield smaller bubbles, mitigating MB coalescence. In the absence of surfactant, the MBG produced MBs of approximately 52 μm. Notably, an augmentation in surfactant concentration boosted gas holdup by entraining more gas and augmented the bubble population by reducing bubble size.

The relationship between salt concentration and bubble size was investigated by Walker et al. (2001), who reported a diminution in bubble size with rise in salt concentration. This reduction is likely attributable to two factors:

first, the decline in surface tension associated with increased salt concentration results in the formation of finer bubbles in the MBG; second, the presence of ions and salt impedes bubble coalescence.

1.4.5 Rheological Properties

MBs find applications in enhanced oil recovery, drug delivery, and in vivo ultrasonic imaging, with surfactant-stabilized MBs being prevalent. However, in water treatment contexts, the avoidance of additives is imperative to prevent further contamination. Some studies indicate that surfactant-stabilized MBs exhibit viscoelastic properties.

In a study conducted by Shams et al. (2017), experiments were undertaken to ascertain the viscosity of MBs within capillary tubes. The value of viscosity of MBs was assessed across three distinct tube lengths and diameters, exploring variations in the void fractions of MBs and bubble size distributions. The MBs exhibited a non-Newtonian behavior, with viscosity decreasing as the Re increased. This decrease persisted until reaching a constant viscosity, beyond which further increases in Re maintained a consistent viscosity of MBs. Notably, the viscosity of MBs diminished with decreasing tube diameter and increasing tube length, attributed to differing bubble orientations within the tube cross-section. Additionally, suspensions' viscosities increased with a rise in bubble void fraction.

Studies state that MB suspension is a non-Newtonian pseudo-plastic fluid, which is time-independent. Its rheology mostly satisfies the power law model, or Ostwald–De Waele model (Parmar and Majumder, 2014; Shen et al., 2008).

$$\tau = K'\left(-\frac{dU_{\mathrm{c}}}{dr}\right)^{n} \tag{1.12}$$

where K' and n are constants dependent on the type of fluid and are independent of the column geometry. U_{c} is the velocity of MBs–particle, and r is the radial distance. The constant K' is denoted as the fluid consistency, and n is the flow behavior index. n and K' can be calculated in a horizontal pipe experimentally from the wall shear stress (τ) and the apparent shear rate (γ_{a}) as (Larmignat et al., 2005):

$$\tau = \alpha\gamma_{a}^{n} \tag{1.13}$$

The parameters α and K' of the MBs–particle mixture are related as (Larmignat et al., 2005):

$$\alpha = K'\left(\frac{3n+1}{4n}\right)^{n} \tag{1.14}$$

The values τ and γ_a are evaluated from the following relation using the volumetric flow rate of mixture (Q_m) and pipe pressure drop (ΔP):

$$\tau = \frac{D_p \Delta P}{4L_p} \tag{1.15}$$

$$\gamma_a = \frac{32Q_m}{\pi D_p^3} \tag{1.16}$$

where D_p is the diameter of pipe and L_p is the length of pipe. The viscosity of MBs–particle mixture (μ_e) in the bubble column was calculated by:

$$\mu_e = K' \left(\frac{8U_c}{D_c} \right)^{n-1} \tag{1.17}$$

Nomenclature

Notation		SI Unit
A	proportionality constant used for Eq. (1.10)	unit less
C_s	concentration of surfactant in the liquid	mg/L
d^{90}, d^{50}, d^{10}	90th, 50th, and 10th volume percentiles of the cumulative size distribution curve	m
d_b	bubble diameter	m
D_b	translational diffusion coefficient of microbubble	m^2/s
d_{BS}	Sauter mean diameter	m
D_c	column diameter	m
di	diameter of ith number bubble	m
D_p	pipe diameter	m
E	electrophoretic mobility	$m^2/V.s$
g	gravitational constant	$9.8\ m/s^2$
K'	fluid consistency constant	unit less
L_p	length of pipe	m
n	flow behavior index constant	unit less
N_a	Avogadro's number	unit less
n_i	number of bubbles in a bubble diameter range of d_i	unit less
P	partial pressure	Pa
P_{atm}	atmospheric pressure	Pa

p_b	pressure inside the bubble	Pa
P_G	gas pressure	Pa
P_L	liquid pressure	Pa
ΔP	pressure difference measured between the two sections	Pa
Q_m	volumetric flow rate of mixture	m^3/s
r	radius of the bubble	m
R	gas constant	J/mol.K
Δs	difference in frame size	s
S	surface area of bubbles	m^2
t	time	s
Δt	time difference	s
T	temperature	K
U_c	velocity of MBs	m/s
V	volume of the bubble	m^3
V_0	rising velocity of the bubble front in the liquid media	1/ms
V_f	final liquid volume	m^3
Vi	initial liquid volume	m^3/s

Greek Nomenclature

γ_a	apparent shear rate	1/s
δ	bubble size dispersion coefficient	
ϵ	dielectric constant of water	unit less
ϵ_L	liquid volume fraction or liquid holdup	unit less
ϵ_0	permittivity of free space	$C^2/J.m$
ζ	Zeta potential	V
μ_e	MBs–particle mixture viscosity	Pa.s
μ_G	viscosity of gas	Pa.s
μ_L	viscosity of liquid (water)	Pa.s
ρ_G	density of gas	kg/m^3
ρ_L	density of water	kg/m^3
σ	the surface tension of the liquid	N/m
τ	wall shear stress	Pa

References

Agarwal, A., Ng, W.J., and Liu, Y. (2010). Principle and applications of microbubble and nanobubble technology for water treatment. Chemosphere 84: 1175–1180.

Ago, K., Nagasawa, K., Takita, J. et al. (2005). Development of an aerobic cultivation system by using a microbubble aeration technology. J. Chem. Eng. Jpn. 38: 757–762.

Agrawal, A., Park, J., Ryu, D.Y. et al. (2005). Controlling the location and spatial extent of nanobubbles using hydrophobically nanopatterned surfaces. Nano Lett. 5: 1751–1756.

Ahmadi, R. and Khodadadi, D.A. (2013). Modelling and optimization of nano-bubble generation process using response surface methodology. Int. J. Nanosci. Nanotechnol. 9: 151–162.

Ahmed, N. and Jameson, G.J. (1985). The effect of bubble size on the rate of flotation of fine particles. Int. J. Miner. Process. 14: 195–215.

Ahmed, A.K.A., Sun, C., Hua, L. et al. (2018). Generation of nanobubbles by ceramic membrane filters: the dependence of bubble size and zeta potential on surface coating, pore size and injected gas pressure. Chemosphere 203: 327–335.

Akimov, V.V., Dmitriev, E.A., and Trushin, A.M. (2011). Mass transfer in the chemosorption of CO_2 in a membrane microbubble apparatus. Theor. Found. Chem. Eng. 45: 811–817.

Akuzawa, H., Amagai, K., Funatsu, M. et al. (2010). Study on cleaning of pipe inner wall by micro-bubble flow. Jpn. J. Multiphase Flow 24: 454–461.

Alam H.S., Soelaiman T.A.F., Soetikno P., Sugiarto A.T. Population balance and computational fluid dynamics modeling of swirl flow microbubble generator. ICSEEA Tangerang Indonesia. 2020; 1–7. https://doi.org/10.1109/ICSEEA50711.2020.9306166.

Alves, S.S., Carvalho, S.P., and Vasconcelos, J.M.T. (2005). Effect of bubble contamination on rise velocity and mass transfer. Chem. Eng. Sci. 60: 1–9.

Badger, S. (2019). *Laser diffraction VS light scattering VS photo-optical analyzing*. W.S Tyler https://blog.wstyler.com/cpa/laser-diffraction-vs-light-scattering-vs-photo-optical-analyzing.

Bando, Y., Takahashi, Y., Luo, W. et al. (2008). Flow characteristics in concurrent up flow bubble column dispersed with micro-bubbles. J. Chem. Eng. Jpn. 41: 562–567.

Berne, B.J. and Pecora, R. (2000). *Dynamic Light Scattering with Applications to Chemistry, Biology, and Physics*. New York: Dover Publications.

Bredwell, M.D. and Worden, R.M. (1998). Mass-transfer properties of micro-bubbles. Biotechnol. Prog. 14: 31–38.

Bui, T.T., Nguyen, D.C., and Han, M. (2019). Average size and zeta potential of nanobubbles in different reagent solutions. J. Nanopart. Res. 21: 1–11.

Burns, S.E., Yiacoumi, S., and Tsouris, C. (1997). Microbubble generation for environmental and industrial separations. Sep. Purif. Technol. 11: 221–232.

Calgaroto, S., Wilberg, K.Q., and Rubio, J. (2014). On the nanobubbles interfacial properties and future applications in flotation. Miner. Eng. 60: 33–40.

Camel, V. and Bermond, A. (1998). The use of ozone and associated oxidation processes in drinking water treatment. Water Res. 32: 3208–3222.

Cavaro, M., Payan, C., Moysan, J., and Baque, F. (2011). Microbubble cloud characterization by nonlinear frequency mixing. J. Acoust. Soc. Am. 129: EL179–EL183.

Chandran, P., Bakshi, S., and Chatterjee, D. (2015). Study on the characteristics of hydrogen bubble formation and its transport during electrolysis of water. Chem. Eng. Sci. 138: 99–109.

Clift, R., Grace, J.R., and Weber, M.E. (1978). *Slow Viscous Flow Past Spheres in Bubbles, drops, and particles*. Academic Press.

Couto, H.J.B., Melo, M.V., and Massarani, G. (2004). Treatment of milk effluent by dissolved air flotation. Braz. J. Chem. Eng. 21: 83–91.

Couto, H., Nunes, D.G., Neumann, R., and França, S. (2009). Micro-bubble size distribution measurements by laser diffraction technique. Miner. Eng. 22: 330–335.

Da Cruz, S.G., Dutra, J.B.A., and Monte, B.M.M. (2016). The influence of some parameters on bubble average diameter in an electroflotation cell by laser diffraction method. J. Environ. Chem. Eng. 4: 3681–3687.

Deendarlianto, W., Tontowi Indarto, A.E., and Iriawan, A.G.W. (2015). The implementation of a developed microbubble generator on the aerobic wastewater treatment. Int. J. Technol. 6: 924–930.

Desai, P., Ng, W.C., Michael, J.H. et al. (2019). Comparison of bubble size distributions inferred from acoustic, optical visualisation, and laser diffraction. Colloids Interfaces 3: 65.

Desai, P., William, B., and Zimmerman. (2023). Transient effects and the role of wetting in microbubble generation. Curr. Opin. Colloid Inter. Sci. 67: 101722.

Dicker, S., Mleczko, M., Hensel, K. et al. (2011). Co-encapsulation of lipid microbubbles within polymer microcapsules for contrast applications. Bubble Sci. Eng. Technol. 3: 12–19.

Edzwald, J.K. (1995). Principles and applications of dissolved air flotation. Water Sci. Technol. 31: 1–23.

Eskanlou, A., Khalesi, M.R., Mirmogaddam, M. et al. (2018). Investigation of trajectory and rise velocity of loaded and bare single bubbles in flotation process using video processing technique. Sep. Sci. Technol. 54: 1–8.

Fan, Y.V., Varbanov, P.S., Klemes, J.J., and Nemet, A. (2018). Process efficiency optimization and integration for cleaner production. J. Clean. Prod. 174: 177–183.

Feshitan, J.A., Chen, C.C., Kwan, J.J., and Borden, M.A. (2009). Microbubble size isolation by differential centrifugation. J. Colloid Interface Sci. 329: 316–324.

Filippov, L.O., Joussemet, R., and Houot, R. (2000). Bubble spargers in column flotation: adaptation to precipitate flotation. Miner. Eng. 13: 37–51.

Ghosh, P. (2009). *Colloid and Interface Science*. PHI Learning Pvt. Ltd.

Glantschnig, W.J. and Chen, S.H. (1981). Light scattering from water droplets in the geometrical optics approximation. Applied Optics 20: 2499–2509.

Gonzalez-Galvis, J.P. and Narbaitz, R.M. (2020). Large batch bench-scale dissolved air flotation system (LB-DAF) for drinking water treatability tests. Environ. Sci. Water Res. Technol. 6: 1004–1017.

Gotoh, Y., Serizawa, A., Eguchi, T. et al. (2006). Oil separation from oil polluted soil by microbubble injection and separation mechanisms. Jpn. J. Multiphase Flow 20: 39–49.

Graciaa, A., Creux, P., Lachaise, J., and Salager, J.L. (2000). Potential at an air-water surface related to the critical micelle concentration of aqueous mixed surfactant systems. Ind. Eng. Chem. Res. 39: 2677–2681.

Grau, R. and Heiskanen, K. (2002). Visual technique for measuring bubble size in flotation machines. Miner. Eng. 15: 507–513.

Gulden, S.J., Riedele, C., Rollie, S. et al. (2018). Online bubble size analysis in micro flotation. Chem. Eng. Sci. 185: 168–181.

Han, M.Y., Park, Y.H., and Yu, T.J. (2002). Development of a new method of measuring bubble size. Water Sci. Technol. 2: 77–83.

Hasegawa, H., Kataoka, H., and Asano, K. (2009). Effect of the electrical potential of microbubbles on ozone dissolution. J. Phys. Confer. Ser. 147: 012024–012032.

Himuro, S., Deguchi, T., and Takamatsu, R. (2009). Effects of microbubbles on bacteria. Prog. Multiphase Flow Res. 4: 95–102.

Hiroshi, T. (2006). Increasing role of ozonation in water purification and sewage treatment. Gekkan Chikyo Kankyo 37: 110–111.

Hofmeier, U., Yaminsky, V.V., and Christenson, H.K. (1995). Observations of solute effects on bubble formation. J. Colloid Interface Sci. 174: 199–210.

Hosokawa, S., Tanaka, K., Tomiyama, A. et al. (2009). Measurement of microbubbles generated by a pressurized dissolution method. J. Phys. Conf. Ser. 147: 012016–012026.

Ikeura, H., Kobayashi, F., and Tamaki, M. (2011). Removal of residual pesticide, fenitrothion, in vegetables by using ozone microbubbles generated by different methods. J. Food Eng. 103: 345–349.

Jauregi, P. and Varley, J. (1998). Colloidal gas aphrons: a novel approach to protein recovery. Biotechnol. Bioeng. 59: 471–481.

Jia, W., Ren, S., and Hu, B. (2013). Effect of water chemistry on zeta potential of air bubbles. Int. J. Electrochem. Sci. 8: 5828–5837.

Jin, F., Li, J., Ye, X., and Wu, C. (2007). Effects of pH and ionic strength on the stability of nanobubbles in aqueous solutions of a-cyclodextrin. J. Phys. Chem. B 111: 11745–11749.

Jyoti, K.K. and Pandit, A.B. (2001). Water disinfection by acoustic and hydrodynamic cavitation. Biochem. Eng. J. 7: 201–212.

Kawahara, A., Sadatomi, M., Matsuyama, F., and Matsuura, H. (2009). Prediction of micro-bubble dissolution characteristics in water and seawater. Exp. Therm. Fluid Sci. 33: 883–894.

Kelsall, G.H., Tang, S., Smith, A.L., and Yurdakul, S. (1996). Measurement of rise and electrophoretic velocities of gas bubbles. J. Chem. Soc. Faraday Trans. 92: 3879.

Kerker, M. (1969). *The scattering of light and other electromagnetic radiation*. New York.: Academic Press.

Khuntia, S., Majumdar, S.K., and Gosh, P. (2012). Microbubble-aided water, and wastewater purification: a review. Rev. Chem. Eng. 28: 191–221.

Khuntia, S., Majumdar, S.K., and Gosh, P. (2013). Removal of ammonia from water by ozone microbubbles. Ind. Eng. Chem. Res. 52: 318–326.

Kim, M.S. and Kwak, D.H. (2017). Effect of zeta potential on collision-attachment coefficient and removal efficiency for dissolved carbon dioxide flotation. Environ. Eng. Sci. 34: 272–280.

Kracht, W., Emery, X., and Paredes, C. (2013). A stochastic approach for measuring bubble size distribution via image analysis. Int. J. Miner. Process. 121: 6–11.

Kukizaki, M. (2009). Microbubble formation using asymmetric shirasu porous glass (SPG) membranes and porous ceramic membranes – a comparative study. Colloids Surf A 340: 20–32.

Kukizaki, M. and Baba, Y. (2008). Effect of surfactant type on microbubble formation behavior using shirasu porous glass (SPG) membranes. Colloids Surf. A 326: 129–137.

Kukizaki, M. and Goto, M. (2006). Size control of nanobubble generated from Shirasu-Porous glass (SPG) membranes. J. Membr. Sci. 281: 386–396.

Kurup, N. and Naik, P. (2010). Microbubbles: a novel delivery system. J. Pharm. Res. Health Care 2: 228–234.

Kwak, D.H., Jung, H.J., Kwon, S.B. et al. (2009). Rise velocity verification of bubble-floc agglomerates using population balance in the DAF process. J. Water Supply Res. Technol. 58: 85.

Larmignat, S., Vanderpool, D., Lai, H.K., and Pilon, L. (2005). Rheology of colloidal gas aphrons (microfoams). Colloids Surf. A Physicochem. Eng. Asp. 322: 199–210.

Lee, J.I., Yim, B.S., and Kim, J.M. (2020). Effect of dissolved-gas concentration on bulk nanobubbles generation using ultrasonication. Sci. Rep. 10: 1–7.

Li P. *Development of advanced water treatment technology using microbubbles.* PhD thesis., School of Science for Open and Environmental Systems, Keio University., Kanagawa, Japan 2006.

Li, X., Li, P., Zu, L., and Yang, C. (2016). Gas-Liquid mass transfer characteristics with microbubble aeration–I. standard stirred tank. Chem. Eng. Technol. 39: 945–952.

Lindner, J.R. (2004). Microbubbles in medical imaging: current applications and future directions. Nat. Rev. Drug Discov. 3: 527–533.

Liu, S., Wang, Q., Ma, H. et al. (2010). Effect of micro-bubbles on coagulation flotation process of dyeing wastewater. Sep. Purif. Technol. 71: 337–346.

Lucero, A. Jr., Kim, D.S., and Park, Y.S. (2017). Parameter optimization for cost reduction of microbubble generation by electrolysis. J. Environ. Sci. Int. 26: 269–280.

Majid A.I., Aan D., Wiratni W., Indarto I. Development of an industrial-scale micro-bubble generator for the purposes of aerobic wastewater treatment. 9th ICMF-2016. Firenze Italy, May 22nd–27th, 2016.

Matsumoto, M. and Tanaka, K. (2008). Nano bubble-Size dependence of surface tension and inside pressure. Fluid Dyn. Res. 40: 546–553.

Meegoda, J.N., Aluthgun, H.S., and Batagoda, J.H. (2018). Stability of nanobubbles. Environ. Eng. Sci. 35: 1216–1227.

Miyamoto, M., Ueyama, S., Hinomoto, N. et al. (2007). Degreasing of solid surfaces by microbubble cleaning. Jap. J. Appl. Phys. 46: 1236–1243.

Nakano, Y., Nishijima, W., Okata, M., and Fukagawa, K. (2005). Microbubble-based groundwater treatment. Groundwater treatment technology with use of activated carbon-based solvent regeneration and ozone. Environ. Solution. Technol. 4: 11–16.

Nazari, S., Shafaei, S.Z., Hassanzadeh, A. et al. (2020). Study of effective parameters on generating submicron (nano)-bubbles using the hydrodynamic cavitation. Physicochem. Probl. Miner. Process. 56: 884–904.

Ohnari H. Swing type fine air bubble generating device. International Patent WO 69550, 2000.

Ohnari H. Swirling fine-bubble generator. US Patent 6382601 B1, 2002.
Onoe, K., Matsumoto, M., and Akiya, T. (2002). Reactive crystallization of calcium carbonate using CO_2 micro-bubbles. Bull. Soc. Sea Water Sci. Jap. 56: 357–361.
Osserman, R. (1978). The isoperimetric inequality. Bull. Amer. Math. Soc. 84: 1182–1234.
Park, J.S. and Kurata, K. (2009). Application of microbubbles to hydroponics solution promotes lettuce growth. Hort Technology 19: 212–215.
Parkinson, L., Sedev, R., Fornasiero, D., and Ralston, J. (2008). The terminal rise velocity of 10–100 μm diameter bubbles in water. J. Colloid Interface Sci. 322: 168–172.
Parmar, R. and Majumder, S.K. (2014). Hydrodynamics of microbubble suspension flow in pipes. Ind. Eng. Chem. Res. 53: 3689–3701.
Pelssers, E.G.M., Stuart, M.A.C., and Fleer, G.J. (1990). Single particle optical sizing (SPOS) I. Design of an improved SPOS instrument and application to stable dispersions. J. Colloid Interface Sci. 137: 350–361.
Qu, X., Wang, L., Karakashev, S.I., and Nguyen, A.V. (2009). Anomalous thickness variation of the foam films stabilized by weak non-ionic surfactants. J. Colloid Interface Sci. 337: 538–547.
Rawle, A. (2002). The importance of particle size to the coating industry part I: particle size measurement. Adv. Col. Sci. Technol. 5: 1–12.
Rehman, F., Medley, F.J.D., Bandulasena, H., and Zimmerman, W.B. (2015). Fluidic oscillator-mediated microbubble generation to provide cost-effective mass transfer and mixing efficiency to the wastewater treatment plants. Environ. Res. 137: 32–39.
Richardson, J.F. and Zaki, W.N. (1997). Sedimentation and fluidization: Part 1. Chem. Eng. Res. Des. 75: S82–S100.
de Rijk, S.E., aivan der Graaf, J.H.J.M., and den Blanken, J.G. (1994). Bubble size in flotation thickening. Water Res. 28: 465–473.
Roy, D., Valsaraj, K.T., and Tamayo, A. (1992). Comparision of soil washing using conventional surfactant solutions and colloidal gas aphron suspensions. Sep. Sci. Technol. 27: 1555–1568.
Sadatomi, M., Kawahara, A., Kano, K., and Ohtomo, A. (2005). Performance of a new micro-bubble generator with a spherical body in a flowing water tube. Exp. Therm. Fluid Sci. 29: 615–623.
Sardeing, R., Painmanakul, P., and Hébrard, G. (2006). Effect of surfactants on liquid side mass transfer coefficients in gas–liquid systems: a first step to modelling. Chem. Eng. Sci. 61: 6249–6260.
Schneider, C.L., Neumann, R., and Souza, A.S. (2007). Determination of the distribution of size of irregularly shaped particles from laser diffractometer measurements. Int. J. Miner. Process. 82: 30–40.

Sebba, F. (1988). *Foams and Biliquid foams (Aphrons)*. New York: John Wiley and Sons.

Shams, M.M., Dong, M., and Mahinpey, N. (2017). Viscosity and rheological behavior of microbubbles in capillary tubes. AIChE J. 60: 2660–2669.

Shangguan, Y., Yu, S., Gong, C. et al. (2018). A review of microbubble and its applications in ozonation. Earth Environ. Sci. 128: 012149.

Shen, Y., Longo, M.L., and Powell, R.L. (2008). Stability and rheological behavior of concentrated monodisperse food emulsifier coated microbubble suspensions. J. Colloid Interface Sci. 327: 204–210.

Sloan, E.D. Jr., (1998). Gas hydrates: review of physical/chemical properties. Energy Fuels 12: 191–196.

Soli, K.W., Yoshizumi, A., Motomatsu, A. et al. (2010). Decontamination of fresh produce by the use of slightly acidic hypochlorous water following pretreatment with sucrose fatty acid ester under microbubble generation. Food Control 21: 1240–1244.

Sumikura, M., Hidaka, M., Murakami, H. et al. (2007). Ozone micro-bubble disinfection method for wastewater reuse system. Water Sci. Technol. 56: 53–61.

Swart, B., Zhao, Y., Khaku, M. et al. (2020). *In situ* characterization of size distribution and rise velocity of microbubbles by high-speed photography. Chem. Engg. Sci. 225: 115836.

Switkes, M. and Ruberti, J.W. (2004). Rapid cryofixation/freeze fracture for the study of nanobubbles at solid–liquid interfaces. Appl. Phys. Lett. 84: 4759–4761.

Takagi, S., Ogasawara, T., and Matsumoto, Y. (2008). The effects of surfactant on the multi-scale structure of bubbly flows. Philos. Trans. R. Soc. Math. Phys. Eng. Sci. 366: 2117–2129.

Takahashi, M. (2005). ζ Potential of microbubbles in aqueous solutions: electrical properties of the gas–water interface. J. Phys. Chem. B 109: 21858–21864.

Takahashi, M. (2009). Base and technological application of micro-bubble and nano-bubble. Mater. Integration 22: 2–19.

Takahashi, M., Chiba, K., and Li, P. (2007). Free-radical generation from collapsing microbubbles in the absence of a dynamic stimulus. J. Phys. Chem. B 111: 1343–1347.

Tasaki, T., Wada, T., Fujimoto, K. et al. (2009). Degradation of methyl orange using short- wavelength UV irradiation with oxygen microbubbles. J. Hazard. Mater. 162: 1103–1110.

Temesgen, T., Bui, T.T., Han, M. et al. (2017). Micro and nanobubble technologies as a new horizon for water-treatment techniques: a review. Adv. Colloid Interface Sci. 246: 40–51.

Terasaka, K. and Shinpo, Y. (2007). Separation of fine particles suspended in water using microbubble flotation. Jpn. J. Multiphase Flow 21: 77–83.

Terasaka, K., Hirabayashi, A., Nishino, T. et al. (2011). Development of microbubble aerator for wastewater treatment using aerobic activated sludge. Chem. Eng. Sci. 66: 3172–3179.

Tsuge, H. (2010). Fundamentals of microbubbles and nanobubbles. Bull. Soc. Sea Water Sci. Jpn. 64: 4–10.

Tsuge, H., Li, P., Shimatani, N. et al. (2009). Fundamental study on disinfection effect of microbubbles. Kagaku Kogaku Ronbunshu 35: 548–552.

Tsutsumi, H. (2010). Application of microbubble injector to marine fish farming and its future perspective. Bull. Soc. Sea Water Sci. Jpn. 64: 31–38.

Ulatowski, K., Jeżak, R., and Sobieszuk, P. (2021). Impact of process parameters on the diameter of nanobubbles generated by electrolysis on platinum-coated titanium electrodes using box–Behnken experimental design. Energies 14: 2542.

Ushikubo, F.Y., Enari, M., Furukawa, T. et al. (2010). Zeta-potential of micro-and/or nano-bubbles in water produced by some kinds of gases. IFAC Proc. 43: 283–288.

Vinnett, L., Sovechles, J., Gomez, C.O., and Waters, K.E. (2018). An image analysis approach to determine average bubble sizes using one-dimensional Fourier analysis. Miner. Eng. 126: 160–166.

Walker, A.B., Tsouris, C., Depaoli, D.W., and Klasson, K.T. (2001). Ozonation of soluble organics in aqueous solutions using microbubbles. Ozone Sci. Eng. 23: 77–87.

Wen, L.H., Ismail, A.B., Menon, P.M. et al. (2011). Case studies of microbubbles in wastewater treatment. Desalin. Water Treat. 30: 10–16.

Wu, C., Nesset, K., Masliyah, J., and Xu, Z. (2012). Generation and characterization of submicron size bubbles. Adv. Colloid Interface Sci. 179: 123–132.

Xiaohui, D., Jingyu, X., Yingxiang, W., and Donghong, L. (2011). Research on the technology of oil removal by dynamic state micro-bubbles flotation. Gongye Shuichuli 31: 89–91.

Xu R. *Particle Characterization: Light Scattering Methods. Particle Technology Series*, 13th Vol., Kluwer Academic Publishers 2002.

Xu, Q., Nakajima, M., Ichikawa, S. et al. (2008). A comparative study of microbubble generation by mechanical agitation and sonication. Innovative Food Sci. Emerg. Technol. 9: 489–494.

Xu, Q., Nakajima, M., Liu, Z., and Shiina, T. (2011). Biosurfactants for microbubble preparation and application. Int. J. Mol. Sci. 12: 462–475.

Yasuda, K., Matsushima, H., and Asakura, Y. (2019). Generation and reduction of bulk nanobubbles by ultrasonic irradiation. Chem. Eng. Sci. 195: 455–461.

Yoon, R.H. and Yordan, J.L. (1986). Zeta-potential measurements on microbubbles generated using various surfactants. J. Colloid Interface Sci. 113: 430–438.

Yu, X., Abe, K., Hirose, Y. et al. (2007). Measurement of solid-liquid two-phase flow using a new robust normal-line Hough transform method. AIP Conf. Proc. 914: 301–306.

Zeng, W., Jia, C., Luo, H. et al. (2020). Microbubble-dominated mass transfer intensification in the process of ammonia-based flue gas desulfurization. Ind. Eng. Chem. Res. 59: 19781–19792.

Zhang, H.B. and Li, J. (2010). Stirring and electric field assisted microbubble preparation. Bubble Sci. Eng. Technol. 2: 9–15.

Zhang, X.H., Li, G., Maeda, N., and Hu, J. (2006a). Removal of induced nanobubbles from water/graphite interfaces by partial degassing. Langmuir 22: 9238–9243.

Zhang, X.H., Maeda, N., and Craig, V.S.J. (2006b). Physical properties of nanobubbles on hydrophobic surfaces in water and aqueous solutions. Langmuir 22: 5025–5035.

2

Mass Transfer of Microbubbles

2.1 Introduction

In the realm of water treatment, the utilization of microbubble (MB) technology stands as an emerging frontier, distinguished by its unparalleled efficiency and quality in comparison to conventional methodologies. This innovative approach to water treatment boasts the capacity to achieve superior results with reduced reliance on coagulants, presenting a marked departure from traditional coagulation practices. MB technology, classified as a form of physical–chemical processing, holds the capability to expedite the intricate processes of flocculation and flotation within water treatment plants (WTPs). This improvement, marked by high-level aeration, emerges as a cornerstone in the optimization of WTPs, enhancing their sustainability and concurrently addressing challenges associated with contaminated raw water sources. The mechanism underlying MB technology, often termed "dissolved air flotation," relies on the principles of bubble adhesion, nucleation, entrapment, and entrainment, wherein bubbles effectively capture floc and transport it to the water surface. Noteworthy is the technology's prowess in accelerating oxidation processes, surpassing the capabilities of conventional aeration units. This heightened efficiency is attributed to MBs' superior mass transfer properties compared to traditional air supply methods. Consequently, MB technology stands poised to meet the escalating demands of clean water treatment, presenting a transformative solution to contemporary challenges.

A pivotal advantage of MB application lies in its ability to enhance the gas mass transfer to the bulk, contributing to a more refined and efficient water treatment paradigm. This technological advancement heralds a new era in the pursuit of water treatment excellence, offering a potent tool for addressing the ever-growing global need for access to safe and clean water. The introduction of fine bubbles, owing to their diminutive size, holds the

Microbubbles: Engineering Aspects and Industrial Applications, First Edition. Snigdha Khuntia.

promise of yielding a substantial specific interfacial area, consequently engendering elevated mass transfer rates. MBs, characterized by their advantageous attributes, provide immense potential for augmenting gas–liquid mass transfer. Enhanced efficiency of mass transfer dynamics surrounding bubbles is attributed to a rise in the efficacy of the gas–liquid contactors, including stirred tanks and bubble columns. Consequently, experimental and simulation studies have been undertaken to unravel the intricacies of macrobubbles possessing conventional sizes ($d_b > 0.1$ mm), given their foundational significance across diverse applications (Griffith, 1960). On the contrary, knowledge concerning the mass transfer of MBs ($d_b \leq 0.1$ mm) remains relatively limited. While numerous studies have explored stationary MBs in the stationary phase (Lozano et al., 2007), the natural mobility of MBs, either rising freely or coursing with the liquid flow, necessitates a deeper understanding of their mass transfer properties during their rise in water. Furthermore, bubbles are commonly found in contaminated environments rather than in pure systems, where several solutes, such as dissolved salts and surfactants, are present in water. Given that these solutes exert a substantial influence on movement and the rate of mass transfer of the MBs, comprehending the impact of these elements becomes essential (Aoki et al., 2015a, 2015b; Hori et al., 2017). Thus, the quest for knowledge extends beyond static conditions, demanding an exploration of the dynamic interplay between rising MBs and the multifaceted compositions characterizing real-world systems.

2.2 Fundamental Mass Transfer Theories Applicable to Microbubbles

2.2.1 Mass Transfer Models

The rate of mass transfer of the gas component i, in MBs represented as $dm_i(t)/dt$, having N number of gases, is articulated by Eq. (2.1). This equation assumes that the individual gas components exhibit independent mass transfer of their own:

$$\frac{dm_i(t)}{dt} = k_{L,i}A(t)(c_{s,i}(t) - c_{b,i}) \tag{2.1}$$

where m denotes the moles of the gas in the MBs and $A = \pi d_b^2$ signifies the MBs' surface area. Henry's law is applied to transform the gas concentration into liquid; Equation (2.1) can be modified to:

$$\frac{dm_i(t)}{dt} = k_{L,i}A(t)[H_i y_i(t) p_b(t) - c_{b,i}] \tag{2.2}$$

where y_i represents the mole fraction of the ith component, calculated by:

$$y_i = \frac{m_i}{\sum_{i=1}^{N} m_i} \tag{2.3}$$

According to the well-known two-film theory of mass transfer proposed by Lewis and Whitman (1924), the overall mass transfer coefficients in both phases, denoted as K_L (mass transfer coefficients at the liquid side) and K_G (mass transfer coefficients at the gas side), are defined by Cussler (1997):

$$\frac{1}{K_L} = \frac{1}{K_G H} = \frac{1}{k_L} + \frac{1}{k_G H} \tag{2.4}$$

The values of Henry's law constant (H) for gases can be found in the literature. For MBs where gas molecules cover a short distance, the gas-phase mass transfer resistance, $1/k_g H$, is negligible. Particularly for lesser soluble gases like O_2 and O_3, the liquid phase predominantly offers the mass transfer resistance (Johnson and Davis, 1996). Consequently, it is reasonable to assume that $K_L \approx k_L$.

The determination of the two-film theory–based mass transfer coefficient, as expressed in Eq. (2.2), poses a formidable challenge. As the gas undergoes dissolution within the liquid phase, its concentration within the liquid experiences a continuous increase. The rate of change in concentration is expressed by a straightforward concentration-based equation, as illustrated in Eq. (2.5):

$$\frac{dc}{dt} = k_L a \left(c_s^* - c_a\right) \tag{2.5}$$

Here, c_s^* and c_a denote the concentrations of saturated gas and concentration of gas over time in the solvent, respectively, and $k_L a$ represents the volumetric mass transfer coefficient. In the majority of gas–liquid contacting apparatus, the precise determination of interfacial area remains elusive. Consequently, the product of k_L and a is employed. This amalgamation, known as the $k_L a$, provides more accuracy. The value of a is intricately linked to ϵ_G, the fractional gas holdup, and d_{BS}, the Sauter mean diameter elucidated by Kawahara et al. (2009):

$$a = \frac{6\epsilon_G}{d_{BS}} \tag{2.6}$$

Assuming an initial gas concentration in water at $t = 0$ is c_0, Equation (2.6) yields the expression of the gas concentration in the water phase:

$$\ln\left(\frac{c_s^* - c_a}{c_s^* - c_0}\right) = -k_L a t \tag{2.7}$$

The value of the $k_L a$ can be calculated experimentally from the Eq. (2.7) by the linear plot of $\ln\left(\frac{c_s^* - c_a}{c_s^* - c_0}\right)$ and t, taking a negative slope of $k_L a$, or by a nonlinear regression analysis.

2.2.2 Methods of Evaluation of Mass Transfer Coefficient

2.2.2.1 From Correlations

The value of k_L can also be calculated using dimensionless numbers, the Reynolds number (Re), the Schmidt number (Sc), and the Sherwood number (Sh), through several correlations. The dimensionless numbers are expressed as follows:

$$\mathrm{Re} = \frac{\rho_L u_b d_b}{\mu_L} \tag{2.8}$$

$$\mathrm{Sc} = \frac{\mu_L}{D \rho_L} \tag{2.9}$$

$$\mathrm{Sh} = \frac{k_L d_b}{D} \tag{2.10}$$

Clift et al. (1978) propose two mass transfer equations for bubbles, considering them as spheres, which have either an immobile or a mobile surface. For these conditions, the Sh is articulated as:

$$\text{For immobile surface, Sh} = 1 + (1 + \mathrm{Pe})^{\frac{1}{3}} \tag{2.11}$$

$$\text{For mobile surface, Sh} = 1 + \left(1 + 0.564\mathrm{Pe}^{\frac{2}{3}}\right)^{\frac{3}{4}} \tag{2.12}$$

where Pe is the Peclet number, signifying the ratio of convective and diffusive mass transfer rates, generally expressed as the product of Re and Sc:

$$\mathrm{Pe} = \mathrm{Re} \times \mathrm{Sc} = \frac{u_b d_b}{D} \tag{2.13}$$

Equations (2.11) and (2.12) are applicable exclusively in the Stokes law region (where $Re < 1$). For a broader applicability, Griffith (1960) introduced a more generic expression:

$$\mathrm{Sh} = \mathrm{Sh}_0 + b\mathrm{Re}^{\frac{1}{2}}\mathrm{Sc}^{\frac{1}{3}} \tag{2.14}$$

where Sh_0 represents the Sherwood number for the stationary sphere, which is under free convection, $\mathrm{Sh}_0 = 2$. Griffith suggests that the coefficient b varies within the range of 0.4 and 0.9. Widely quoted and utilized, the correlation from Ranz and Marshall (1952) is presented as:

$$\mathrm{Sh} = 2 + 0.6\mathrm{Re}^{\frac{1}{2}}\mathrm{Sc}^{\frac{1}{3}} \tag{2.15}$$

This correlation is applicable at a no-slip condition at the gas–liquid interface, in the range of $0 < \text{Re} < 200$, and valid only when the bubble surface is completely contaminated. The applicability of these formulations is evaluated through a comparative analysis with experimentally determined values.

Sato et al. (1981) originally suggested that the turbulence of the bubbles is directly proportional to the product of the d_b and the bubble rising velocity, u_G. To validate this hypothesis, the data for k_L was plotted against the product of the d_{BS} and the u_G. The results indicated a unique relationship between k_L and $u_G d_{BS}$. Subsequently, a dimensionless form of k_L, incorporating surface tension (σ) and liquid viscosity (μ_L), was introduced. Additionally, the data were correlated using a straightforward equation:

$$\frac{k_L \mu_L}{\sigma} = 7.46 \times 10^{-10} \text{Pe}^{0.756} \tag{2.16}$$

It is essential to conduct further investigations to assess the suitability of Eq. (2.16) for liquids with properties divergent from those of water and seawater.

2.2.2.2 From Comparison of the Saturation of Gases

One common approach for determining the mass transfer coefficient of a gas–liquid system involves utilizing Henry's law constant. In a meticulously degassed or air-bubbled sample with a volume of 0.5 L, the concentration of dissolved oxygen c_{O_2} is measured both before and after the introduction of MBs. The average values of c_{O_2} before and after the generation of MBs, observations are taken and employed to compute the air saturation ratio (S_a) as:

$$S_a = \frac{c_{O_2}}{c^*_{O_2}} \tag{2.17}$$

where $c^*_{O_2}$ is the saturated oxygen concentration dissolved in water at temperature T. From the experimental data of c_{O_2}, the concentration of dissolved N_2 (c_{N_2}) is calculated by the assumption of an equivalent saturation ratio attained for c_{N_2} as that for c_{O_2}. This assumption can be expressed as:

$$S_a = \frac{c_{N_2}}{c^*_{N_2}} = \frac{c_{O_2}}{c^*_{O_2}} \tag{2.18}$$

This assumption has been accepted and tested by various studies on mass transfer from single stationary MBs (Duncan and Needham, 2004; Kwan and Borden, 2010). The saturation gas concentration is determined by considering Henry's law, as follows:

$$c^*_i = H_i p_i = H_i y_i P_{atm} \tag{2.19}$$

The evaluation of the mass transfer coefficient, k_L, is based on the assessment of the bubble diameter rate of change, d_b. Due to the presence of noise in bubble diameter measurements, moving averages were employed over a specific certain time interval ranging from 0.2 to 0.5 seconds. This interval was determined based on the rate of bubble shrinkage and recording the frame rate.

$$n = \frac{p_b V}{RT} = \frac{p_b \pi d_b^3}{6RT} \tag{2.20}$$

where p_b represents the pressure inside the bubble in Pa. The rearrangement of Eq. (2.1) results in the expression for the k_L, as:

$$k_L = \frac{1}{A(c_s - c_b)} \frac{dn}{dt} \tag{2.21}$$

The instantaneous k_L for O_2 can be determined using Eq. (2.18) assuming the mole fractions of the gaseous component (y) (such as of O_2 and N_2) remain constant during the process of dissolution, i.e. $y_{O_2} = 0.21$ and $y_{N_2} = 0.79$.

$$k_{L,O_2} = \frac{y_{O_2}}{A(c_{s,O_2} - c_{b,O_2})} \frac{dn}{dt} \tag{2.22}$$

Subsequently, the k_L value can be evaluated by substituting the surface area $A = \pi d_b^2$, and the surface concentration $c_{s,O_2} = H_{O_2} y_{O_2} p_b$ into Eq. (2.23). A centered difference between t_1 and t_2 and for $\frac{dn}{dt}$ is applied as follows:

$$k_{L,O_2} = \frac{y_{O_2}}{6RT(d_{b,12})^2 (H_{O_2} y_{O_2} p_{b,12} - c_{b,O_2})(c_{s,O_2} - c_{b,O_2})} \frac{p_{b,2} d_{b,2}^3 - p_{b,1} d_{b,1}^3}{t_2 - t_1} \tag{2.23}$$

where subscripts 1, 2, and 12 denote the values at t_1, t_2, and $t_{12} = (t_1 + t_2)/_2$, respectively, and $d_{b,12}$ and $p_{b,12}$ are the arithmetic average of the values at t_1 and t_2, respectively. The values of d_b with times t_1 and t_2 are experimentally obtained. This method for evaluation of k_L was originally proposed and used by Hosoda et al. (2014) and later by Kastens et al. (2015).

2.2.2.3 Using the Plug Flow Model

Bredwell and Worden (1998) developed a methodology utilizing a mathematical model to predict the variations in bubble diameter in the widely used bubble columns, enabling the separate estimation of the k_L and a. High-density dispersed MBs were generated through a spinning disk apparatus initially proposed by Sebba (1988). The apparatus consists of a high-speed electric motor to rotate a 5 cm diameter, having a thickness of

1 cm, at a rotational speed above 4000 rpm. Localized high shear was created by stationary baffles positioned within 5 mm of the spinning disk. The disk and baffles were of stainless steel and were fixed in a 6 L fermentation vessel. The setup was connected to provide the rotational speed with the motor attached. The MB sample was recirculated through a pump and diluted in water to achieve a concentration of 7–15%. The data were recorded roughly for 60 s. The nonionic Tween20 surfactant known for its MB-forming ability and biocompatibility with *Butyribacterium methylotrophicum*, the target fermenter compound, was utilized. The Tween20 concentration of 120 mg/L, double the critical micelle concentration (CMC), was employed. Surfactant-assisted MBs were generated with 100% oxygen.

The model proposed by Bredwell and Worden (1998) was under the assumption that the tracer compound (usually a salt) only stays in the liquid phase and remains uniform throughout the radial direction. Also, the tracer peak corresponding to each probe underwent normalization through a calibration curve. For an open vessel, the unsteady-state mass balance of the tracer compound in the liquid phase is expressed as follows:

$$\frac{\partial c_T}{\partial t} = D_{ax}\frac{\partial^2 c_T}{\partial z^2} - u_l\frac{\partial c_T}{\partial z} \tag{2.24}$$

where c_L represents the liquid-phase oxygen concentration (mg/L). The relevant boundary conditions are as follows:

$$\begin{array}{lll} t = 0 & z = 0 & c_L = c_0 \\ t > 0 & z = 0 & c_L = 0 \\ t < 0 & z = \infty & c_L = 0 \end{array}$$

To evaluate the optimum values for the dispersion coefficient, a computer program was originally developed by Wakao and Kaguei (1982); however, Thompson and Worden (1997) modified it later. This program minimized the least squares error by the conversion of the experimental values into Fourier series and calculating a predicted tracer curve from Eq. (2.24). The gas phase was assumed to travel in plug flow throughout the column, a presumption generally suitable for systems with a substantial length-to-diameter (l/D_c) ratio. In the case of the bubble column used by Bredwell and Worden (1998), the length-to-diameter ratio was 24.

The plug flow model also helps determine the value of k_La as follows. Assuming plug flow, the provided steady-state mass balance equation at the liquid phase for oxygen is expressed as follows:

$$\frac{dc_L}{dz} - \frac{k_La}{u_l}\left(c_L^* - c_L\right) = 0 \tag{2.25}$$

With the boundary conditions, $z = 0$, and $c_L = 0$.

The value of a can be expressed as:

$$a = \frac{3r^2 \epsilon_G}{r_0^3} \tag{2.26}$$

where the value of ϵ_G varies with the bubble size according to:

$$\epsilon_G = \epsilon_{G,0} \left(\frac{r}{r_0} \right)^3 \tag{2.27}$$

The concentration of oxygen at the interface changes with atmospheric pressure, depth of liquid in the column, and bubble size, as shown below:

$$c_L^* = H \left[\rho \left(\frac{g}{g_c} \right) (l - z) + P_{atm} + \frac{2\sigma}{r} \right] \tag{2.28}$$

The rate of shrinkage of MBs varies significantly across the column height and is expressed by the following equation:

$$\frac{dr}{dz} = \frac{\rho \left(\frac{g}{g_c} \right) r - \frac{3k_L}{u} \left(c_L^* - c_L \right) \frac{RT}{M}}{3P_{atm} + 2(l - z)\rho \left(\frac{g}{g_c} \right) + \frac{4\sigma}{R}} \tag{2.29}$$

with the boundary condition at $z = 0$ and $r = r_0$.

Equation (2.29) considers the changes that happen across the bubble column due to the difference in hydrostatic pressure of MB, Laplace pressure, and mass transfer occurring due to the loss of gas. Simusolv modeling software was used for solving Eqs. (2.25)–(2.29) (Dow Chemical Co., Midland, MI) employing the gear stiffness method, considering a reduced-gradient optimization on k_L. The average r value produced by the MB generator (MBG) was determined using the Dynamic Light Scattering (DLS) method. In addition, the initial value of a was calculated with respect to the flow rates of the water and MB dispersion.

Another important parameter calculation, the dispersion coefficient, D_{ax}, can be predicted from this model. The dispersion coefficients for MBs in the bubble column are depicted and calculated, as shown in Eq. (2.30). Additionally, the correlation captured from the literature in a conventional bubble column was included for comparative purposes:

$$D_{ax} = 2.7 d_{tube}^{1.4} u_G^{0.3} \tag{2.30}$$

It was observed that the axial dispersion for MB dispersions is notably less than that for usual bubble dispersions. In traditional systems, macrobubbles rise at a significant velocity as compared to the surrounding liquid, creating wakes that transport liquid through the height of the column at the higher velocity, thus contributing to the axial mixing. The velocity of slow-rising

MBs eliminates the formation of such wakes. The values of D_{axial} were then utilized to calculate Pe numbers, determined to be 42 at the lower side, indicative of lower axial mixing. The plug flow mixing model was suitable for calculating k_L values from the oxygen-profile data for bubble columns.

2.2.2.4 Using Higbie's, and Frössling's Theories

For a situation in the MB-filled bubble columns, where the system is considered to have the mobile gas–liquid interface, due to the less contact time of the gas–liquid in the pond, the penetration of the gas phase is less. In this situation, Higbie's theory (Higbie 1935) is known to be valid, shown as:

$$k_{L,\text{Higbie}} = \sqrt{\frac{4D_L u_r}{\pi d_{BS}}} \tag{2.31}$$

where u_r is the bubble–liquid slip or rise velocity and D_L is the molecular diffusion coefficient of the liquid phase. On the contrary, in the case of a rigid interface, Frössling's equation, originally derived from laminar boundary layer theory, is known to be better applicable:

$$k_{L,\text{Frössling}} = c\sqrt{\frac{u_b}{d_{BS}}} D_L^{2/3} \nu_L^{-1/6} \tag{2.32}$$

where c is a constant, excrementally approximated as 0.6, and ν_L is the liquid kinetic viscosity. The obtained k_L values increase with the size of the bubbles but decrease with increasing the concentration of salt. The increase in k_L might be due to a rise in the bubble-induced turbulence, which is directly proportional to the size of the bubble.

Nevertheless, a contrary trend between k_L and bubble size emerged when comparing the calculations with the data reported by Kawahara et al. (2009). Consequently, modifications were necessary to adapt Higbie's and Frössling's equations to align with the present dataset. To address this, the ratios of experimental to calculated values were introduced:

$$f_H = k_{L,(\text{experimental})} / k_{L,\text{Higbie}} \tag{2.33}$$

$$f_F = k_{L,(\text{experimental})} / k_{L,\text{Frössling}} \tag{2.34}$$

Here, $k_{L,(\text{experimental})}$ represents the experimental k_L; $k_{L,\text{Higbie}}$ and $k_{L,\text{Frössling}}$ denote the calculated k_L using Eqs. (2.33) and (2.34), respectively; and f_H and f_F were plotted against the Ohnesorge number, Z:

$$Z = \frac{\mu_L}{(\rho_L d_{BS} \sigma)^{0.5}} = \left(\frac{\text{Mo}}{\text{Eo}}\right)^{0.25} \tag{2.35}$$

$$\text{Eo} = \frac{g(\rho_L - \rho_G) d_{BS}^2}{\sigma} \tag{2.36}$$

$$\mathrm{Mo} = \frac{g\mu_L^4(\rho_L - \rho_G)}{\rho_L^2\sigma^3} \tag{2.37}$$

where Eo is the Eötvös number and Mo is the Morton number. The f_H and f_F were established as follows:

$$f_H = 1.10 \times 10^{-7} Z^{-2.74} \tag{2.38}$$

$$f_F = 6.87 \times 10^{-7} Z^{-2.71} \tag{2.39}$$

In conclusion, k_L can be determined using the following equations:

$$k_L = 1.10 \times 10^{-7} Z^{-2.74} \sqrt{\frac{4D_L u_B}{\pi d_{BS}}} \tag{2.40}$$

$$k_L = 6.87 \times 10^{-7} Z^{-2.71} c \sqrt{\frac{4u_B}{d_{BS}}} D_L^{2/3} \nu_L^{-1/6} \tag{2.41}$$

Nedeltchev et al. (2006) introduced a modified $k_L a$ correlation, derived from an adaptation of Higbie's theory (1935):

$$k_L a = f_c \sqrt{\frac{4D_L u_G}{\pi d_{BS}} \frac{6V_G H_L}{V_L d_{BS} u_G}} \tag{2.42}$$

In this equation, V_G represents the volumetric gas flow rate, and H_L is the height of the liquid phase in the column. The value of f_c is evaluated using a correction factor proposed by Nedeltchev et al. (2006).

$$f_c = 0.185 \mathrm{E}_o^{0.737} \tag{2.43}$$

A thorough experiment was carried out to calculate and propose Eq. (2.43) based on data from 14 organic liquids and tap water, covering an Eo from 1.7 to 7. This range corresponds to situations where the bubble shape deviates from a spherical form, exhibiting an ellipsoidal geometry.

2.3 Dissolution Behavior of Microbubbles

MBs exhibit essential characteristics such as higher gas dissolution, minimal rising velocity, and a significant interfacial area. The paramount factor influencing MBs' properties is the bubble size distribution. Narrow bubble size distribution enhances the gas–liquid interfacial area, leading to increased gas retention and a subsequent decrease in bubble rise velocity. Another crucial parameter to consider is k_L of oxygen or k_{L,O_2}, reflecting the gas dissolution behavior in water. In the case of MB-based aeration systems, a combination of specially designed MBG and fluid flow rates is employed to achieve a high k_{L,O_2}. Both experimental and computational investigations indicate that an increase in the Re amplifies the impact force on MBs, resulting in a reduction in bubble diameter and an extended retention time of MBs in the water.

Figure 2.1 illustrates the dissolution characteristics of an MB in ultrapure water under 100% air saturation, or $S = 1$. The experiments were depicted in a photographic series in Figure 2.1 (a), which visually captures the dissolution process, while Figure 2.1 (b) depicts the time-dependent progression of the bubble diameter during its shrinkage. The dissolution time, denoted as τ, represents the moment when the bubble completely vanished from the captured images. As the image processing capabilities in MATLAB have a limitation in diameter of approximately 10 μm, τ was determined through visual inspection. Here, the time points $t = 0$ and $t = t_c$ indicate the moments of liquid injection and the clear appearance of a bubble in focus within the high-speed camera's field of view, respectively.

The value of $k_L a$ is highly dependent on the U_G (superficial gas velocity) and ϵ_G. In a study conducted by Kawahara et al. (2009), three types of test liquids, such as tap water, 1% seawater, and 3% seawater, were considered to investigate the effects of salinity on MBs dissolution behavior. Figures 2.2 a and b illustrate the values obtained for $k_L a$ for the three liquids with change in U_G, and ϵ_G. Irrespective of the test liquids, $k_L a$ exhibits a linear rise with both U_G and ϵ_G. At a fixed U_G, $k_L a$ increases with the concentration of salt, while at a fixed ϵ_G, $k_L a$ decreases with concentration. To comprehend these dissolution characteristics, $k_L a$ is further dissected into two parameters, k_L and a, separately.

The value of U_G is inversely proportional to the salt concentration due to a reduction in d_{BS}, owing to decreased surface tension and the reduced bubble coalescence. On the contrary, $k_L a$ increases with an increase in salt concentration. Ultimately, the substantial increase in a at high salt concentrations emerges as the primary factor behind the elevated $k_L a$ values.

To assess the efficiency of oxygen absorption, E_A was employed, representing the ratio of the rate of oxygen absorption to the rate of oxygen supply in the liquid, calculated as:

$$E_A = \frac{N_C}{\dot{m}_{O_2}} \times 100\% \tag{2.44}$$

Here, $\dot{m}_{O_2}$ denotes the mass flow rate of O_2, if from air, at 23% of the air supply rate. N_c represents the mass transfer rate of O_2, defined by:

$$N_C = k_L a V_L c_s \tag{2.45}$$

Here, V_L is the volume of liquid in the bubble column. E_A decreases with a rising superficial air velocity when $U_G < 0.0005$ m/s. E_A is directly proportional to the salt concentration caused by the prolonged residence time of the MBs in the liquid and the higher interfacial area as compared to the tap water.

2.4 Effect of Parameters on k_L

2.4.1 Saturation Ratio

Figure 2.3 represents the rate of MBs shrinkage, $\frac{dd_b}{dt}$, in ultrapure water, as a function of saturation ratio, ranging from 0.2 to 1. It also shows the impact of degassing, represented by the saturation ratio, S, on the shrinkage rate. A discernible trend emerges, with the MB shrinkage rate escalating as the saturation ratio decreases. The plot in Figure 2.3 demonstrates that MBs approximately 30 μm in diameter exhibit a consistent shrinkage rate

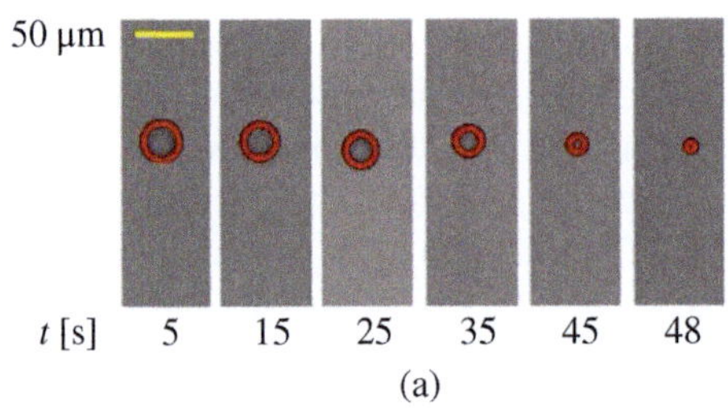

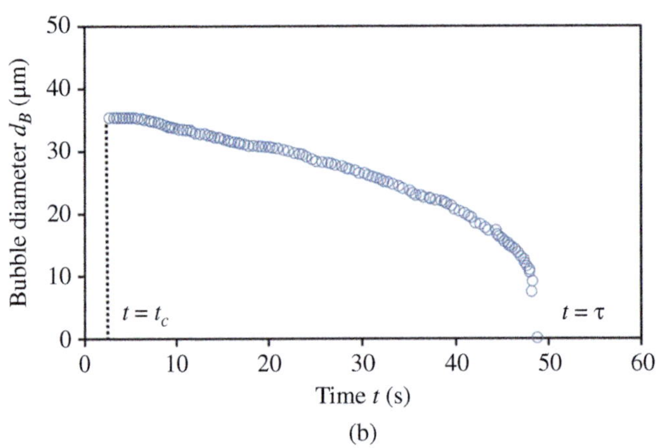

Figure 2.1 The dissolution of MBs in water (a) captured in a high-speed camera at different times for a single bubble dissolution and (b) the variation of bubble diameter during the dissolution process. Source: Tanaka et al. (2019)/ with permission of ELSEVIER.

contingent on the value of S, reflecting the degree of the driving force. But, for bubbles down to 30 μm size, the shrinkage rate gradually intensifies across all examined saturation ratios. For example, a Laplace pressure of 5.7 kPa for d_b = 50 μm, contrasting with 14.4 kPa for d_b = 20 μm, assuming the surface tension of water, σ = 72 mN/m. This indicates that bubbles of 20 μm size possess a driving force for mass transfer more than two times greater than those with a diameter of 50 μm. Consequently, it underscores the necessity to consider the influence of surfactants on surface tension when modeling mass transfer properties for MBs of smaller diameter.

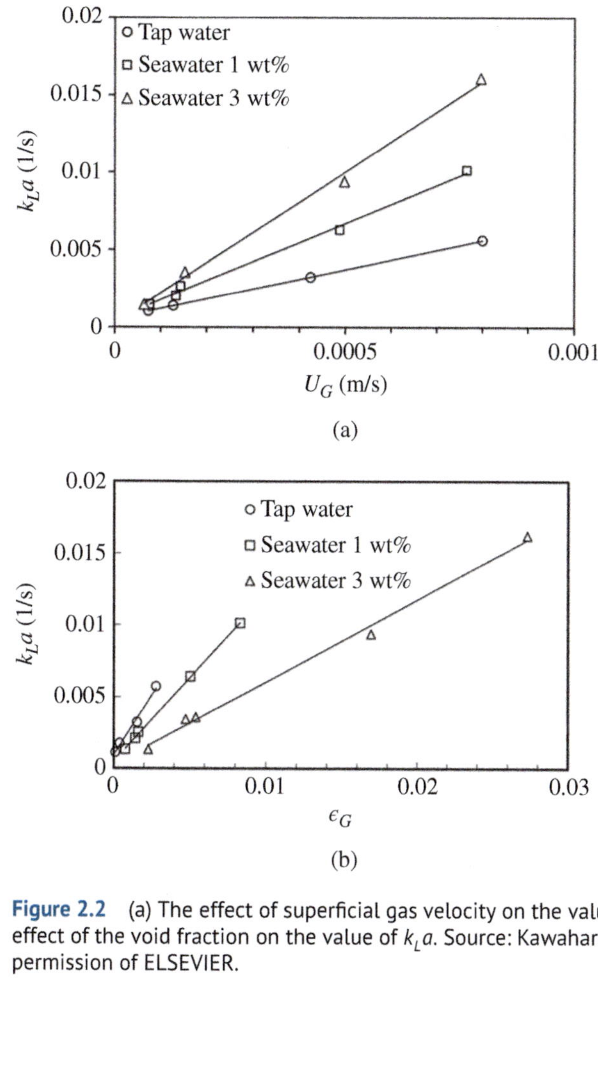

Figure 2.2 (a) The effect of superficial gas velocity on the value of k_La and (b) the effect of the void fraction on the value of k_La. Source: Kawahara et al. (2009)/with permission of ELSEVIER.

2.4.2 Presence of Surfactants

In many applications, surfactants play a pivotal role in influencing the ascent speed, coalescence, and mass transfer processes (Fahad et al., 2019). Alves et al. (2005) have explored the mass transfer characteristics of individual bubbles within the 1–5 mm diameter range, employing the stagnant cap model to interpret the influence of contaminants. It is observed by researchers that k_L values remain unchanged, approximately 1×10^{-4} m/s for bubbles lower than 1.5 mm, regardless of the contamination status of the aqueous liquid phase (Painmanakul et al., 2005; Sardeing et al., 2006).

Aoki et al. (2015a, 2015b) explored the nature of bubbles of diverse shapes, ranging from ellipsoidal to elongated (also known as Taylor bubbles). Utilizing numerical simulations to study the distribution of the surfactants along the surface of the bubble, they uncovered that the transition from a clean, mobile surface to a contaminated, immobile surface resulted in a reduction in mass transfer, a phenomenon attributed to the Marangoni effect. As the size of the bubble decreases, the immobilization of the bubble surface accelerates owing to surfactant adsorption. This conclusion finds support in numerical investigations of relatively smaller spherical bubbles (Cuenot et al., 1997; Takemura, 2005). Additionally, these numerical studies highlighted that surfactants exhibit more homogeneous adsorption on the bubble surface ($d_b < 0.5$ mm), characterized by a low concentration gradient, as compared to macrobubbles. The studies further estimated that it takes less than 0.5 seconds for the surfactants to get fully immobilized on the surface of MBs ($d_b \leq 0.1$ mm). Consequently, it is assumed that surfactants rapidly

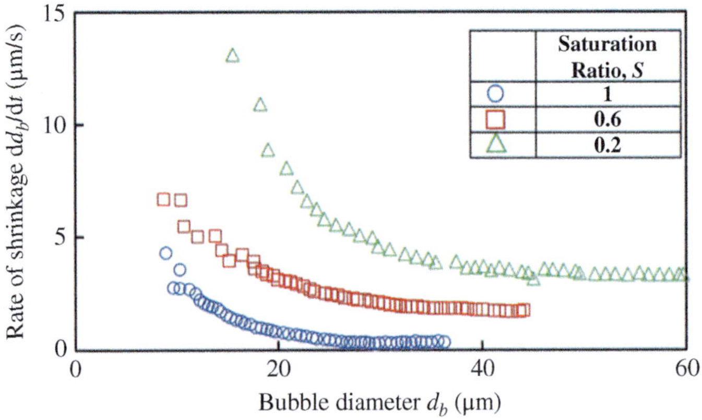

Figure 2.3 The rate of MB shrinkage affected by the saturation ratio. Source: Tanaka et al., 2019/with permission of ELSEVIER.

adsorb onto MB surfaces during the process of bubble generation, independent of the concentration of the surfactant.

Surfactants also play a dual role in reducing the surface tension and consequently affecting the rate of mass transfer of MBs, as indicated by the Laplace equation $\Delta P = \frac{4\sigma}{d_b}$. However, the molecular weight of the surfactants can significantly influence the mass transfer properties of MBs, introducing the concept of "shell resistance" (Bredwell and Worden, 1998; Kwan and Borden, 2010; Salari et al., 2017) or "diffusion barrier" (Shim et al., 2014). In the study reported by Tanaka et al. (2019), the impact of molecular weight is deemed almost negligible for low concentrations of surfactant. This suggests that the decrease in mass transfer rate and longer dissolution time primarily result from the decreased surface tension, specifically the reduced mass transfer driving force related to the Laplace pressure.

2.4.3 Presence of Salts

The most common salt, sodium chloride has a negligible impact on surface tension at the gas–liquid interface but does influence the solubility of other solutes as well as the diffusion coefficient, a phenomenon referred to as the salt effect. Therefore, the study of salt effect on diffusion coefficients and Henry's constants is considered in the calculation of dissolution processes of MBs. Figure 2.4 illustrates the influence of NaCl concentration, C_{NaCl}, on the dissolution of MBs under air, at $S = 1$, utilizing diffusion coefficients and Henry's constants. The figure indicates that higher salt concentrations result in longer dissolution times, with the predicted value of dissolution

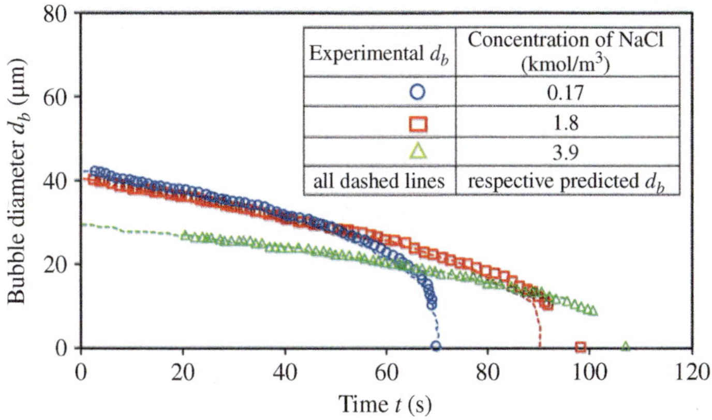

Figure 2.4 The behavior of dissolution of MBs at different salt concentrations at $S = 1$. Source: Tanaka et al. (2019)/with permission of ELSEVIER.

closely aligning with experimental results. This implies that the presence of salt decreases the rate of mass transfer largely through two mechanisms: the decline in Henry's constants, leading to a declining driving force, and the decrease in gaseous diffusion coefficients, impacting the mass transfer coefficient.

2.5 Dimensional Analysis of k_La

In the context of a swirl-type MBG, Marwani et al. (2022) introduced a dynamic physical absorption model, employing it to determine the k_La for oxygen. A comparable approach can be adopted for other MBGs, facilitating the subsequent calculation of k_La. Among the key factors influencing k_La, the bubble diameter stands out as a primary parameter. The design features of the MBG, including nozzle diameter and water flow rate, not only impact bubble size but also exert an influence on k_La. Recognizing this, Marwani et al. (2022) conducted a comprehensive examination of parameters affecting the average d_b through dimensional analysis. These parameters are intricately linked to the dimensional analysis of k_La, providing valuable insights into the intricacies of mass transfer processes in MBGs.

Based on the experimental trends, the function describing the average d_b is expressed by Eq. (2.46):

$$d_b = f(u_l, u_G, d_o, D_G, D_{FC}, d_{ch}, \omega, \mu_L, \rho_L, \sigma, \rho_G, \ell, L) \tag{2.46}$$

The famous Buckingham's pi theorem yields 11 pi parameters (π), which are dimensionless, with repeating variables such as u_l, d_o, and L. The nondimensional group to represent d_b is empirically correlated, as shown in Eq. (2.47):

$$\frac{d_b}{d_o} = a_1\left(\frac{u_l^2\rho_L D_{FC}}{\sigma}\right)^{b_1}\left(\frac{u_G}{u}\right)^{c_1}\left(\frac{\omega D_{FC}}{V_{out}}\right)^{d_1}\left(\frac{\ell}{L}\right)^{e_1}\left(\frac{D_G}{D_{FC}}\right)^{f_1}\left(\frac{\rho_G}{\rho_L}\right)^{g_1}$$
$$\left(\frac{D_{FC}}{d_{ch}}\right)^{h_1}\left(\frac{D_G}{d_{ch}}\right)^{i_1}\left(\frac{d_{ch}}{d_o}\right)^{j_1}\left(\frac{d_b}{D_{FC}}\right)^{k_1}\left(\frac{X}{d_o}\right)^{l_1} \tag{2.47}$$

Here, a_1–l_1 represent the coefficients. For swirl-type MBG, specific values of $\ell = 1$, 5, and 10 m were considered. Additionally, Q_L = water flow rate, Q_G = air flow rate, γ = fluid flow velocity ratio, and $\mathrm{We_L}$ = Weber number for water are formulated in subsequent equations:

$$\mathrm{We_L} = \left(\frac{u_l^2\rho_L D_{FC}}{\sigma}\right) \tag{2.48}$$

$$\gamma = \frac{\omega D_{FC}}{V_{out}} \tag{2.49}$$

$$Q_G = \frac{4u_G}{\pi d_o^2} \tag{2.50}$$

$$Q_L = \frac{4u_l}{\pi D_{FC}^2} \tag{2.51}$$

Substituting Eqs. (2.48)–(2.51) in Eq. (2.47) yields:

$$\frac{d_b}{d_o} = a_1(We_L)^{b_1}\left(\frac{Q_G}{Q_L}\right)^{c_1}(\gamma)^{d_1}\left(\frac{\ell}{L}\right)^{e_1}\left(\frac{D_G}{D_{FC}}\right)^{f_1}$$
$$\left(\frac{\rho_G}{\rho_L}\right)^{g_1}\left(\frac{D_{FC}}{d_{ch}}\right)^{h_1}\left(\frac{D_G}{d_{ch}}\right)^{i_1}\left(\frac{d_{ch}}{d_o}\right)^{j_1}\left(\frac{d_b}{D_{FC}}\right)^{k_1} \tag{2.52}$$

The variables D_{FC}, D_G, d_o, and d_{ch} representing dimensions of the MBG, are generally held constant due to infrequent changes. Similarly, ρ_L and ρ_g are physical fluid parameters. Consequently, Eq. (2.52) transforms into Eq. (2.53):

$$\frac{d_b}{d_o} = a_1(We_L)^{b_1}\left(\frac{Q_G}{Q_L}\right)^{c_1}(\gamma)^{d_1}\left(\frac{\ell}{L}\right)^{e_1} \tag{2.53}$$

The coefficients a_1–e_1 were determined by aligning experimental observations, leading to the proposal of the empirical correlation for predicting the average bubble diameter:

$$\frac{d_b}{d_o} = 0.031(We_L)^{-0.015}\left(\frac{Q_G}{Q_L}\right)^{0.082}(\gamma)^{-0.015}\left(\frac{\ell}{L}\right)^{0.014} \tag{2.54}$$

Additionally, considering the impact of temperature on $k_L a$, a correction is applied at $T = 20\,°C$ using the formula by Tchobanoglous et al. (2003):

$$k_L a_{(20)} = k_L a\, 1.024^{(20-T)} \tag{2.55}$$

No significant difference in $k_L a$ is observed until $T = 30\,°C$. Expressing the functional relationship between $k_L a$ and various parameters, Eq. (2.56) is derived:

$$\frac{k_L a d_o^2}{D_{FC}} = a_1\left(\frac{u_l^2\rho_L D_{FC}}{\sigma}\right)^{b_1}\left(\frac{u_G}{u}\right)^{c_1}\left(\frac{\omega D_{FC}}{V_{out}}\right)^{d_1}\left(\frac{\ell}{L}\right)^{e_1}\left(\frac{D_G}{D_{FC}}\right)^{f_1}$$
$$\left(\frac{\rho_G}{\rho_L}\right)^{g_1}\left(\frac{D_{FC}}{d_{ch}}\right)^{h_1}\left(\frac{D_G}{d_{ch}}\right)^{i_1}\left(\frac{d_{ch}}{d_o}\right)^{j_1}\left(\frac{d_b}{D_{FC}}\right)^{k_1}\left(\frac{X}{d_o}\right)^{l_1} \tag{2.56}$$

Simplifying for constant variables, Eq. (2.56) becomes:

$$\frac{k_L a d_o^2}{D_{FC}} = a_1\left(\frac{u_l^2\rho_L D_{FC}}{\sigma}\right)^{b_1}\left(\frac{u_G}{u_l}\right)^{c_1}\left(\frac{\omega D_{FC}}{V_{out}}\right)^{d_1}\left(\frac{\ell}{L}\right)^{e_1}\left(\frac{X}{d_o}\right)^{l_1} \tag{2.57}$$

Through a trial-and-error approach, coefficients a_1–e_1 and l_1 were determined, resulting in the modification of Eq. (2.57) to:

$$\frac{k_L a d_o^2}{D_U} = 0.001(\mathrm{We_L})^{0.474}\left(\frac{Q_G}{Q_L}\right)^{-0.223}(\gamma)^{5.651}\left(\frac{\ell}{L}\right)^{-0.202}\left(\frac{X}{d_o}\right)^{-0.738} \tag{2.58}$$

Many conclusions arise from the above correlation. The rise of the surface tension does not decrease the $k_L a$ much to the notice. Kukizaki and Goto (2006) and Juwana et al. (2019) reported that d_b increases with surface tension; hence, the bubble contact area per unit volume against the liquid decreases. It was found that surface tension has no significant effect on the d_b. Therefore, $\mathrm{We_L}$ has also a negligible effect on $k_L a$ and can be excluded from Eq. (2.58). The value of γ has a positive impact on the $k_L a$, whereas the increase of the turbulent flow and tangential velocity will significantly increase the $k_L a$. Park and Yang (2017) and Juwana et al. (2019) reported that the mass transfer resistance at the interface declines due to the increase of turbulence and the generation of smaller bubbles by the swirling flow. The gap between the air nozzle and the MBG outlet (denoted as ℓ) has a directly proportional relation with $k_L a$.

2.6 Mass Transfer in Reactive System

The mass transfer in a reactive system is more complicated than a simple dissolution process in MBs. The presence of a reactive gas in the MB will not only dissolve in water but also get involved in the reaction of the pollutants or any other compound of particular interest. If the gas phase in the MB is O_3, then the rate of disappearance of O_3 is dependent on two major factors, such as self-decomposition of O_3 and rate of reaction of O_3 with other compounds present in the medium. The total mass balance of O_3 in a system without any other compound can be expressed in Eq. (2.59), considering a continuous O_3 flow rate into the system (a semi-batch process). The concentration of O_3 will attain a constant after a certain time, denoted as saturated O_3 concentration, $[O_3^s]$ (Khuntia et al., 2013, 2014, 2020a, 2020b):

$$\frac{d[O_3]}{dt} = k_L a\left\{[O_3^s] - [O_3]\right\} - k_d[O_3] \tag{2.59}$$

However, in the presence of a pollutant or other reactants, O_3 will be consumed simultaneously for the oxidation process, and Eq. (2.59) will be changed to:

$$\frac{d[O_3]}{dt} = k_L a\left\{[O_3^s] - [O_3]\right\} - k_d[O_3] - k_1[X]^m[O_3]^n \tag{2.60}$$

The disappearance of a compound X reacting with O_3 can be written as:

$$\frac{d[X]}{dt} = -k_1[X]^m[O_3]^n \tag{2.61}$$

where k_d is the self-O_3 decomposition rate, which is dependent on the pH of water. The values of k_d range from 2.33×10^{-4} 1/s to 1.33×10^{-3} 1/s at pH 5–9 (Sotelo et al., 1987). k_1 is the rate constant for the corresponding reaction of O_3 and the compound, and m and n are the orders of the respective reactants. If the rate of reaction of the compound X with O_3 is known, k_La can be calculated through two steps. First, the O_3 concentration with respect to time is to be measured, and second, solving Eqs. (2.60) and (2.61) simultaneously to get the value of k_La. The concentration profile of O_3 and X with respect to time has to be matched for the experimental and the model values, as shown in Figure 2.5.

The most commonly used reaction for a gas–liquid mixture is expressed as:

$$aA + bB \xrightarrow{k_2} \text{Products} \tag{2.62}$$

where A is the reactant in the gas phase and B is the reactant dissolved in the liquid phase. This condition is widely used for water treatment where the pollutants (component B) are treated with O_3 through macrobubbles and MBs (Dalal et al., 2023).

The involvement of Hatta number, Ha, and enhancement factor, E_a, in the reactive mass transfer depicts a vital role in the comparison of the rate of mass transfer and rate of reaction. Ha is expressed as the ratio of the rate

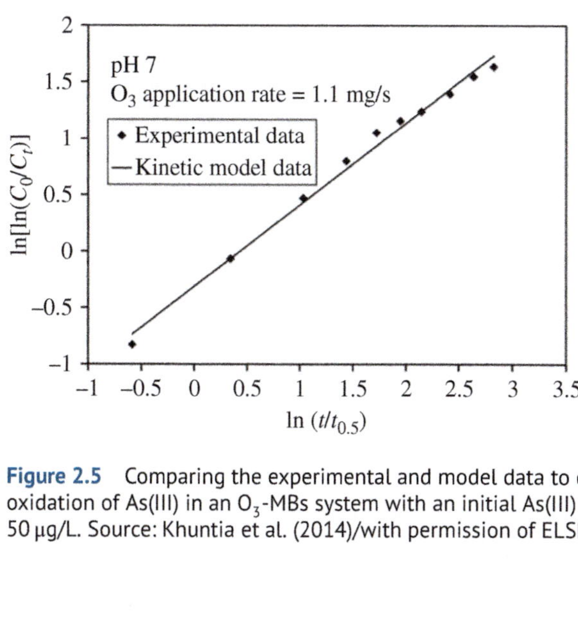

Figure 2.5 Comparing the experimental and model data to evaluate k_La in the oxidation of As(III) in an O_3-MBs system with an initial As(III) concentration of 50 µg/L. Source: Khuntia et al. (2014)/with permission of ELSEVIER.

of chemical reaction and mass transfer in a liquid film. In the MB system, it is often required to find the dominating factor to understand the reaction regimes. *Ha* can be simply expressed as Eq. (2.63):

$$Ha^2 = \frac{\text{Maximum conversion in the liquid} - \text{film}}{\text{Maxium mass transfer diffusion through gas} - \text{film}} \tag{2.63}$$

$$Ha = \frac{\sqrt{kDC_B}}{k_L} \tag{2.64}$$

where k is the rate constant of the reaction, which could take place either at the bulk or at the gas–liquid interface, and D is the diffusion coefficient of the gas present in the MBs in the liquid phase. For a second-order reaction, the *Ha* is simplified to Eq. (2.64). C_B is the concentration of component B in the bulk liquid. The value of *Ha* is a deciding factor to conclude that the reaction completely occurs in the bulk phase if $Ha < 1$, but if $Ha > 1$, then the reaction occurs in the gas–liquid interface (Khuntia et al., 2021). The values of *Ha* also describe the significance of the MBs in the reactive mass transfer system, and a higher value of *Ha* is considered to be the advantage of using MBs. Since the MBs provide a high interfacial area, the usual *Ha* in the reaction medium in the MBs columns is always >1. The commonly used based for MB system and their properties are shown in Table 2.1.

In the case of a reactive gas like O_3, the rate of absorption of O_3 in water can be even more, if O_3 also reacts with the solute (or pollutant) in the bulk phase. This phenomenon is defined as the enhancement factor E_a, which is mostly defined as:

$$E_a = \frac{\text{Rate of mass transfer of the gas in a reacting condition}}{\text{Rate of mass transfer of the gas in non} - \text{reacting condition}} \tag{2.65}$$

Table 2.1 Physical properties of gases used for MBs at 298 K (in water medium).

Gas phase	*H* (mol/m³ Pa)	*D* (m²/s)
CO_2	3.3×10^{-4}	2.2×10^{-9}
N_2	6.4×10^{-6}	2.0×10^{-9}
O_2	1.2×10^{-5}	2.3×10^{-9}
O_3	1.2×10^{-4}	1.76×10^{-9}
Cl_2	9.2×10^{-4}	1.89×10^{-9}

Source: Shane et al. (2014), Sander (2015); Berry et al. (2017).

Table 2.2 The relationship of type of reaction, Hatta number (*Ha*) and enhancement factor (E_a), and their physical meaning.

Type of reaction	Situation at the bulk or interface	*Ha*	E_a	The physical meaning in the situation
Very slow reaction	No reaction in the film, only bulk phase reaction dominating	<0.02	<1	Requirement of a very high liquid holdup
Slow reaction	Abundantly bulk phase reaction	0.02–0.3	<1	Both interfacial area and liquid holdup that should be high
Fast reaction (in the diffusion film)	Reaction happened completely in the interface	>3	$Ha = E_a$	Rate of gas absorption being directly proportional to the interfacial area, the dissolved gas reacting completely at the interface
Fast pseudo-first-order reaction	$[B]_{\text{bulk}} \gg [A]^*$	$3 < Ha < E_a/2$		Prevalence for a pseudo-first-order reaction
Instantaneous reaction	Very fast reaction, rate of reaction being determined by diffusion alone	$>10\ E_a$		For dilute amount of B in the bulk, excess gas dissolution, too high interfacial area

Note: $[A]^*$ is the saturation concentration of the gas component in the liquid and $[B]_{\text{bulk}}$ is the concentration of solute (or pollutant) in the liquid.

A comprehensive relation of the *Ha* and E_a for different types of reactions has been summarized in Table 2.2.

Nomenaclature

Notation		SI unit
A	bubble surface area	m^2
a	interfacial area per unit volume	m^2/m^3

Notation		SI unit
$c_{N_2}^*$	saturated nitrogen concentration	mol/L
c_{N_2}	nitrogen concentration	mol/L
c_{O_2}	oxygen concentration dissolved in water	mol/L
c_L^*	saturated concentration in the liquid phase	mol/L
c_L	concentration in the liquid phase	mol/L
c_{O2}^*	saturated oxygen concentration in dissolved water	mol/L
c_s^*	concentrations of saturated gas	mol/L
c_s^*	saturation concentration of gas in the solvent	mol/L
c_0	initial gas concentration	mol/L
c_a	concentration of gas over time in the solvent	mol/L
c_b	concentration of gas at the bulk phase	mol/L
c_s	the dissolved gas concentration at the gas–liquid interface	mol/L
c_T	concentration of tracer (calcium chloride, Bredwell and Worden [1998]) in liquid phase	mg/L
D	gas diffusion coefficient	m^2/s
D_{ax}	axial dispersion coefficient	m^2/s
d_b	bubble diameter	m
D_b	translational diffusion coefficient of microbubble	m^2/s
d_{BS}	Sauter mean diameter	m
d_{ch}	Chamber diameter	m
D_{FC}	fluid channel diameter	m
D_G	air inlet diameter	m
D_L	liquid-phase molecular diffusion coefficient	m^2/s
dm	difference in moles of the gas	unit less
d_o	MBG outlet diameter	m
d_{tube}	tube diameter	m
D_U	gas–liquid mass diffusivity	m^2/s
E_A	efficiency of oxygen absorption	%
E_a	enhancement factor	unit less
fc	correction factor	unit less
f_F	Frössling's constant	unit less
f_H	Higbie's constant	unit less

Notation		SI unit
g	gravitational constant	9.8 m/s^2
g_c	gravitational conversion factor	unit less
H	Henry's constant	$mol/m^3.Pa$
Ha	Hatta number	unit less
k_1	rate constant for the corresponding reaction of O_3 and the compound	concentra-tion$^{1\text{-order}}$/time
k_2	rate constant for the target compound and the oxidant/reactant	concentra-tion$^{1\text{-order}}$/time
k_d	self-O_3 decomposition rate	1/s
K_G	overall gas-side mass transfer coefficient	mol/N.s
k_G	gas phase mass transfer coefficient	mol/N.s
k_L	liquid phase mass transfer coefficient	m/s
K_L	overall liquid-side mass transfer coefficient,	m/s
l	column length	m
L	chamber length	m
$\dot{m}_{O_2}$	mass flow rate of oxygen supplied as bubbles in the liquid	kg/s
m	number of moles of the gas component in the microbubble	unit less
M	molecular weight	g/mol
n	number of moles of gas	unit less
N_c	oxygen mass transfer rate	mol/s
$[O_3^s]$	saturation concentration of O_3 in water	mol/L
$[O_3]$	concentration of O_3 in water	mol/L
P	partial pressure	Pa
P_{atm}	atmospheric pressure	Pa
p_b	pressure inside the bubble	Pa
P_G	gas pressure	Pa
p_i	vapor pressure of ith component	Pa
P_L	liquid pressure	Pa
Q_G	air flow rate	m^3/s
Q_L	water flow rate	m^3/s
R	gas constant	J/mol.K
r	radius of the bubble	m

Notation		SI unit
r_0	initial bubble radius	m
S_a	saturation ratio	unit less
T	temperature	K
t	time	s
u_b	bubble rising velocity	m/s
u_G	gas velocity	m/s
U_G	superficial gas velocity	m/s
u_l	liquid velocity	m/s
u_r	Bubble liquid slip velocity	m/s
V	volume of bubble	m^3
V_L	volume of liquid in the bubble column	m^3
V_{out}	outlet water velocity	m/s
y	mole fraction of the gas component in the gas phase	unit less
z	position in column	m

Green Nomenclature

Notation		SI unit
$\varepsilon_{G,0}$	initial gas void fraction	unit less
ε_G	gas void fraction	unit less
ℓ	air nozzle gap to MBG outlet	m
γ	fluid flow velocity ratio	unit less
μ_G	viscosity of gas	Pa.s
μ_L	viscosity of liquid (water)	Pa.s
ν_L	kinematic viscosity of liquid	m^2/s
ρ_G	density of gas	kg/m^3
ρ_L	density of water	kg/m^3
σ	surface tension of the liquid	N/m
ω	rotation speed	1/s

Dimensionless Numbers

Notation		SI unit
E_O	Eötvös number	$\frac{g\left(\rho_L - \rho_G\right) d_{BS}^2}{\sigma}$
Mo	Morton number	$\frac{g\mu_L^4\left(\rho_L - \rho_G\right)}{\rho_L^2\sigma^3}$
Pe	Peclet number	$Re \times \mathrm{Sc} = \frac{u_b d_b}{D}$
Re	Reynolds number	$\frac{\rho_L u_b d_b}{\mu_L}$
Sc	Schmidt number	$\frac{\mu_L}{D\rho_L}$
Sh	Sherwood number	$\frac{k_L d_b}{D}$
We_L	Weber number	$\left(\frac{u_l^2 \rho_L D_{FC}}{\sigma}\right)$
Z	Ohnesorge number	$\frac{\mu_L}{\left(\rho_L d_{BS}\sigma\right)^{0.5}} = \left(\frac{M_o}{E_O}\right)^{0.25}$

References

Alves, S.S., Carvalho, S.P., and Vasconcelos, J.M.T. (2005). Effect of bubble contamination on rise velocity and mass transfer. Chem. Eng. Sci. 60: 1–9.

Aoki, J., Hayashi, K., Hosokawa, S., and Tomiyama, A. (2015a). Effects of surfactants on mass transfer from single carbon dioxide bubbles in vertical pipes. Chem. Eng. Technol. 38: 1955–1964.

Aoki, J., Hayashi, K., Hosokawa, S., and Tomiyama, A. (2015b). Mass transfer from single carbon dioxide bubbles in contaminated water in a vertical pipe. Int. J. Heat Mass Transf. 83: 652–658.

Berry, M.J., Taylor, C.M., King, W. et al. (2017). Modelling of ozone mass-transfer through non-porous membranes for water treatment. Water 9: 452.

Bredwell, M.D. and Worden, R.M. (1998). Mass-transfer properties of micro-bubbles. Biotechnol. Prog. 14: 31–38.

Clift, R., Grace, J.R., and Weber, M.E. (1978). *Slow Viscous Flow Past Spheres in Bubbles, Drops, and Particles*. Academic Press.

Cuenot, B., Magnaudet, J., and Spennato, B. (1997). The effects of slightly soluble surfactants on the flow around a spherical bubble. J. Fluid Mech. 339: 25–53.

Cussler, E.L. (1997). *Diffusion Mass Transfer in Fluid Systems*. Cambridge, UK: Cambridge University Press.

Dalal, P., Dalai, S., and Khuntia, S. (2023). Experimental study of the removal NO_x and SO_2 from flue gas using the O_3/H_2O_2, and O_3/UV processes. Environ. Tech. https://doi.org/10.1080/09593330.2023.2256991.

Duncan, P.B. and Needham, D. (2004). Test of the Epstein–Plesset model for gas microparticle dissolution in aqueous media: effect of surface tension and gas undersaturation in solution. Langmuir 20: 2567–2578.

Fahad, M.K., Prakash, R., Majumder, S.K., and Ghosh, P. (2019). Gas holdup in the gas–liquid–coal slurry flow in a flotation column in presence of surface-active agent. Multiphase Sci. Technol. 31: 199–214.

Griffith, R.M. (1960). Mass transfer from drops and bubbles. Chem. Eng. Sci. 12: 198–213.

Higbie, R. (1935). The rate of absorption of a pure gas into a still liquid during short periods of exposure. Trans. Am. Inst. Chem. Engrs. 31: 365–389.

Hori, Y., Hayashi, K., Hosokawa, S., and Tomiyama, A. (2017). Mass transfer from single carbon- dioxide bubbles in electrolyte aqueous solutions in vertical pipes. Int. J. Heat Mass Transf. 115: 663–671.

Hosoda, S., Abe, S., Hosokawa, S., and Tomiyama, A. (2014). Mass transfer from a bubble in a vertical pipe. Int. J. Heat Mass Transf. 69: 215–222.

Johnson, P.N. and Davis, R.A. (1996). Diffusivity of ozone in water. J. Chem. Eng. Data 41: 1485–1487.

Juwana, W.E., Widyatama, A., Dinaryanto, O. et al. (2019). Hydrodynamic characteristics of the microbubble dissolution in liquid using orifice type microbubble generator. Chem. Eng. Res. Des. 141: 436–448.

Kastens, S., Hosoda, S., Schlüter, M., and Tomiyama, A. (2015). Mass transfer from single Taylor bubbles in mini channels. Chem. Eng. Technol. 38: 1925–1932.

Kawahara, A., Sadatomi, M., Matsuyama, F., and Matsuura, H. (2009). Prediction of micro-bubble dissolution characteristics in water and seawater. Exp. Therm. Fluid Sci. 33: 883–894.

Khuntia, S., Majumder, S.K., and Ghosh, P. (2014). Oxidation of As(III) to As(V) using ozone microbubbles. Chemosphere 97: 120–124.

Khuntia, S., Majumder, S.K., and Ghosh, P. (2013). Removal of ammonia from water by ozone microbubbles. Ind. Eng. Chem. Res. 52: 318–326.

Khuntia, S., Sinha, M.K., and Singh, P. (2021). Theoretical and experimental investigation of the mechanism of the catalytic ozonation process by using a manganese-based catalyst. Environ. Technol. 42: 632–639.

Khuntia, S., Sinha, M.K., and Mohan, G. (2020a). Evaluation of reaction kinetics for removal of NO_x by ozone and hydrogen peroxide. Ind. Eng. Chem. Res. 59: 17806–17814.

Khuntia, S., Sinha, M.K., and Saini, B. (2020b). Conversion of NO_2 through ozonation and peroxone process in gas and aqueous phase: Finding the suitable process through experimental route. Chem. Eng. J. 387: 124082.

Kukizaki, M. and Goto, M. (2006). Size control of nanobubble generated From Shirasu-Porous Glass (SPG) membranes. J. Membr. Sci. 281: 386–396.

Kwan, J.J. and Borden, M.A. (2010). Microbubble dissolution in a multigas environment. Langmuir 26: 6542–6548.

Lewis, W.K. and Whitman, W.G. (1924). Principles of gas absorption. Ind. Eng. Chem 16: 1215–1220.

Lozano, M.M., Talu, E., and Longo, M.L. (2007). Dissolution of microbubbles generated in seawater obtained offshore: Behaviour and surface tension measurements. J. Geophys. Res. 112: C12001.

Mawarni, I.D., Juwana, E.W., Yuana, A.K. et al. (2022). Hydrodynamic characteristics of the microbubble dissolution in liquid using the swirl flow type of microbubble generator. J. Water. Proces. Eng. 48: 102846.

Nedeltchev, S., Jordan, U., and Schumpe, A. (2006). A new correction factor for theoretical prediction of mass transfer coefficients in bubble columns. J. Chem. Eng. Jpn. 39: 1237–1242.

Painmanakul, P., Loubière, K., Hébrard, G. et al. (2005). Effect of surfactant on liquid-side mass transfer coefficients. Chem. Eng. Sci. 60: 6480–6491.

Park, S.K. and Yang, H.C. (2017). Experimental investigation on mixed jet and mass transfer characteristics of horizontal aeration process. Int. J. Heat Mass Transf. 113: 544–555.

Ranz, W.E. and Marshall, W.R. (1952). Evaporation from drops. Chem. Eng. Prog. 48: 141–146.

Salari, A., Gnyawali, V., Griffiths, I.M. et al. (2017). Shrinking microbubbles with microfluidics: mathematical modelling to control microbubble sizes. J. Soft Matter 13: 8796–8806.

Sander, R. (2015). Compilation of Henry's law constants (version 4.0) for water as solvent. Atmos. Chem. Phys. 15: 4399–4981.

Sardeing, R., Painmanakul, P., and Hébrard, G. (2006). Effect of surfactants on liquid-side mass transfer coefficients in gas–liquid systems: a first step to modelling. Chem. Eng. Sci. 6: 6249–6260.

Sato, Y., Sadatomi, M., and Sekoguchi, K. (1981). Momentum and heat transfer in two-phase bubble flow-I. Theory. Int J. Multiphase Flow 7: 167–177.

Sebba, F. (1988). *Foams and Biliquid Foams (Aphrons)*. New York: John Wiley and Sons.

Shane, P., Cadogan, G.C., Maitland, J.P., and Trusler, M. (2014). Diffusion coefficients of CO_2 and N_2 in water at temperatures between 298.15 K and 423.15 K at pressures up to 45 MPa. Chem. Eng. Data 59: 519–525.

Shim, S., Wan, J., Hilgenfeldt, S. et al. (2014). Dissolution without disappearing: multicomponent gas exchange for CO_2 bubbles in a microfluidic channel. Lab Chip 14: 2428.

Sotelo, J.L., Beltrán, F.J., Benítez, F.J., and Beltrán-Heredia, J. (1987). Ozone decomposition in water: kinetic study. Ind. Eng. Chem. Res. 26: 39–43.

Takemura, F. (2005). Adsorption of surfactants onto the surface of a spherical rising bubble and its effect on the terminal velocity of the bubble. Phys. Fluids. 17: 048104.

Tanaka, S., Kastens, S., Fujioka, S. et al. (2019). Mass transfer from freely rising microbubbles in aqueous solutions of surfactant or salt. J. Chem. Eng. 387: 121246.

Tchobanoglous, H., Burton, F.L., and Stensel. (2003). *Wastewater Engineering: Treatment and Reuse*, 4the. New York: Mc Graw Hill.

Thompson, V.S. and Worden, R.M. (1997). Phase holdup, liquid dispersion, and gas-to-liquid mass transfer measurements in a three-phase magneto fluidized bed. Chem. Eng. Sci. 52: 279–295.

Wakao, N. and Kaguei, S. (1982). Heat and Mass Transfer in Packed Beds. New York: Gordon and Breach.

3

Hydrodynamics of Microbubbles

3.1 Introduction

The microbubbles (MBs) are nowadays under a lot of attention for their utilization in the multiphase reactors for varied applications. Multiphase reactors are being constantly modified and used for bioreactors, petrochemical applications, sewage treatment, and nonreactive systems such as mineral processing and beneficiations. MBs are proven to be an easy technology for their simple operation and advantages over macrobubbles. A correct design of the reactors and columns required for these operations is crucial. Traditionally, the fluid flow pattern or the parameters of the fluid regime in the bubble columns and multiphase reactors are measured through combined intrusive and nonintrusive methods, usually through conductivity probes, differential pressure transmitters, electrical resistance tomography (ERT), fiberoptic probes, digital photography, etc. (Ning et al., 2020).

The most useful hydrodynamic properties of the bubbles are the rising velocity and residence time, drag coefficient, and gas holdup, which are useful for calculating the mass transfer coefficient in the reacting and nonreacting mediums (Khuntia et al., 2013). In this chapter, the fundamentals of MB hydrodynamics and their methods of calculations have been discussed.

3.2 Hydrodynamic Parameters

3.2.1 Bubble Rising Velocity

The rise of MBs in a liquid is assessed by the equilibrium achieved between buoyancy and drag forces. In water, where the Reynolds number (Re) for

Microbubbles: Engineering Aspects and Industrial Applications, First Edition. Snigdha Khuntia.

MBs remains considerably below unity, the velocity to rise, or "rising velocity" (u_b), is described by the following expression:

$$u_b = \frac{d_b^2 g(\rho_L - \rho_G)}{18\mu_L} \tag{3.1}$$

This equation holds true only for spherical bubbles of identical size in a laminar flow, which in reality is challenging to achieve under practical circumstances (Shen et al., 2008). The rising velocity is intricately linked to the physical properties of the surrounding liquids and the diameter of the bubbles. Due to their relatively small size, MBs tend to exhibit a slower rising velocity, resulting in significantly longer periods of suspension beneath the water surface in comparison to traditional, larger bubbles. Empirical data, as reported in the literature (Takahashi, 2005), closely adheres to Stokes' law, as demonstrated in Figure 3.1.

Bubbles are frequently characterized by their shape and flow behavior, often expressed in terms of dimensionless numbers. These dimensionless numbers help us understand key factors governing bubble dynamics. The Re indicates the comparison of inertial force and the viscous force, while the Eötvös number (Eo) represents the balance between the gravitational force and surface tension of the medium. Additionally, Morton number (Mo) is typically used in combination with Eo to find the shape of the rising bubbles. While predicting the shape of the bubbles in the column, an invaluable tool is the Grace diagram, which takes into account these dimensionless

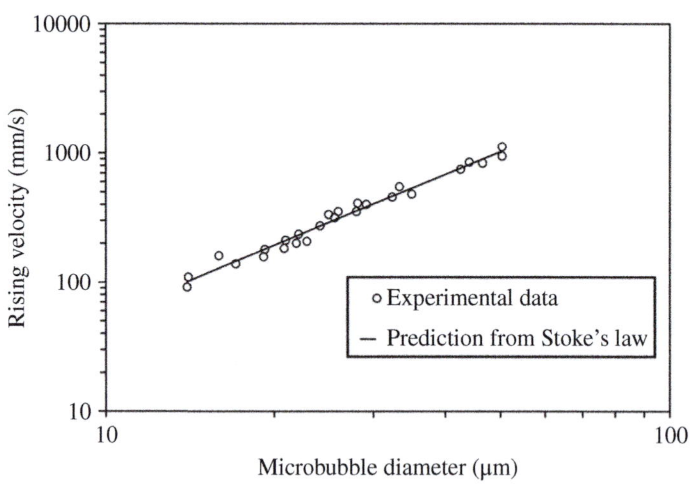

Figure 3.1 The MBs rising speed in distilled water with electrical conductivity below 1.5 μS/cm. Source: Adapted from Takahashi (2005).

numbers together. The expression of Re, Eo, and Mo in context with the MBs has been mentioned in Chapter 2 (Clift et al., 1978).

Macrobubbles rise at an accelerated rate owing to their lower surface-to-volume ratio, resulting in diminished drag and consequently a higher Re. Regardless of the Mo, bubbles tend to maintain a spherical shape when subjected to gravitational forces within a fluid, especially when Re is less than 1 or when the Eo is below 0.2. For instance, an air-MB or air bubble in water with a diameter D_b less than 1.2 mm, typically retains its spherical form. When the surface tension of the medium is sufficiently high, bubbles can remain in spherical shape even with Re_b below 600. In situations where the bubbles exhibit minimal internal circulation, the bubble's behavior is somehow similar to the solid spherical particles. Various correlations exist for approximating the bubble drag coefficient, which acts like solid spheres, aligning with the standard drag curve. However, in the presence of internal circulation, the hydrodynamics of bubbles rising through the liquid in the column become more complex and often require numerical methods for accurate analysis.

For bubbles having $Re_b < 1$, the flow pattern is called the creeping flow around the bubble. In such cases, Stokes' law (3.1) serves as a suitable model for describing the rising speed of an isolated bubble. In that case, the modified Hadamard–Rybczynski (H–R) expression becomes applicable for mobile bubble surfaces considering the internal circulation (Clift et al. 1978).

$$u_{b(H-R)} = \frac{d_b^2(\rho_L - \rho_G)g(\mu_l + \mu_g)}{6\mu_L(2\mu_L + 2\mu_g)} \tag{3.2}$$

3.2.2 Bubble Drag Coefficient

The movement of bubbles within low Re fluids is primarily influenced by factors such as viscosity, surface tension, and inertia (Haapala et al., 2010). When bubbles deviate from the spherical shape and start oscillating, the water surface tension acts as a dominant force in governing their motion. In cases where there is a constant bubble shape, as in the case of MBs, larger bubbles within the same medium tend to ascend more rapidly, unless any change in their motion is hindered by collisions and/or attachments to solid particles dispersed in the flow. The spherical immobile interface bubble drag coefficient in a static pure liquid is estimated as a function of the Re_b using the following correlations. Equation (3.3) is proposed by Hadamard (1911) and Equation (3.4) by Moore (1963).

$$C_D = \frac{16}{\mathrm{Re}_b} \tag{3.3}$$

where $\mathrm{Re}_b < 11$

$$C_D = \frac{48}{\mathrm{Re}_b}\left(1 - \frac{2.21}{\sqrt{\mathrm{Re}_b}}\right) \tag{3.4}$$

where $11 < \mathrm{Re}_b \leq 500$
where

$$\mathrm{Re}_b = \frac{d_b|\vec{u}_b - \vec{u}_l|}{\nu_l} \tag{3.5}$$

$$C_D = \frac{16}{\mathrm{Re}_b}\left\{1 + \left[\frac{8}{\mathrm{Re}_b} + \frac{1}{2}\left(1 + 3.315\mathrm{Re}_b^{-\frac{1}{2}}\right)\right]^{-1}\right\} \tag{3.6}$$

Mei et al. (1994) have established an empirical correlation to determine the drag coefficient for the clean bubbles (free of surfactants or any pollutants) [Eq. (3.6)] that satisfies both correlations Eqs. (3.3) and (3.4), making it applicable across the entire spectrum of Re_b values:

Another criterion used for the prediction of the behavior of MBs in the bubble column is the Bond criterion. It states that for $\mathrm{Eo} < 4$, the internal circulation of gas is negligible for the rising bubbles. Table 3.1 shows a summary of various applicable expressions of drag coefficients for spherical particles at $\mathrm{Re}_b > 1$ by Clift et al. (1978).

Table 3.1 The recommended correlations of drag coefficients at different Re values.

Correlation	Range	Equation number
$C_{DST} = \frac{24}{Re_b}$	$Re_b < 1$, Stoke's law	(3.7)
$C_{DST} = \frac{24}{Re_b}\left[1 + \frac{3}{16}Re_b\right]$	$Re_b < 0.1$, Oseen's approximation	(3.8)
$\left(\frac{C_D}{C_{DST}} - 1\right) = \frac{3c}{16}Re_b, c = 0.43$	$Re_b < 0.1$, Modified Oseen's approximation	(3.9)
$\log_{10}\left[\frac{C_D Re_b}{24} - 1\right] = -0.881 + 0.82w - 0.05w^2$	$0.01 < Re_b \leq 20$	(3.10)
$\log_{10}\left[\frac{C_D Re_b}{24} - 1\right] = -0.7133 + 0.6305w$	$20 \leq Re_b \leq 260$	(3.11)
$\log_{10} C_D = -1.1242w + 0.6305w^2$	$260 \leq Re_b \leq 1500$	(3.12)

3.2.3 Gas Holdup

The term, gas holdup in any bubble column is defined as the fractional volume of the gas phase occupied by bubbles as compared to the volume of the liquid. In an MB column, the gas holdup, ϵ_G, is determined using the phase-isolation technique using the following equation (Karamah et al., 2010).

$$\epsilon_G = \left(1 - \frac{\rho_m}{\rho_L}\right) \tag{3.13}$$

In MB systems, the liquid volume undergoes only minimal changes before and after the introduction of bubbles. Consequently, the precision of the phase-isolation technique diminishes as both the size of MBs and the volume of liquid being tested decrease significantly. To measure gas holdup, the electrical conductivity method was employed in the literature (Parmar and Majumder, 2014). Another technique to measure the ϵ_G in bubble columns was developed by Maxwell (1892). He noted that the effective conductivity of a dispersion denoted as $k_{l\text{-}d}$ is correlated with the ϵ_G of a nonconductive dispersed phase and has expressed the correlation using the equation.

$$k_{l-d} = k_l\left(\frac{1-\epsilon_d}{1+0.5\epsilon_d}\right) \tag{3.14}$$

Based on this principle, gas holdup can be calculated as

$$\epsilon_G = k_l\left(\frac{k_l - k_{l-g}}{k_l + 0.5k_{l-g}}\right) \tag{3.15}$$

k_l is the liquid electrical conductivity, and k_{l-g} represents liquid–gas mixture electrical conductivity. Figure 3.2 illustrates the schematic of the equipment used for the measurement of gas holdup (Kasper, 1940, Parmar and Majumder, 2014). The electrical conductivity offered by the two electrodes can be described using the equation:

$$K_l = \frac{I}{V_t} = k_l\frac{A_e}{L_e} \tag{3.16}$$

where A_e/L_e is denoted as the cell constant. This type of cell has been employed for gauging the effective conductivities of dispersions. To prevent electrode polarization, alternating current (AC) at a relatively high frequency (approximately 1000 Hz) and low voltage (around 1.5 V) was utilized (Parmar and Majumder, 2014). There are a few other available methods used for the measurement of gas holdup, discussed as follows (Haapala et al., 2010):

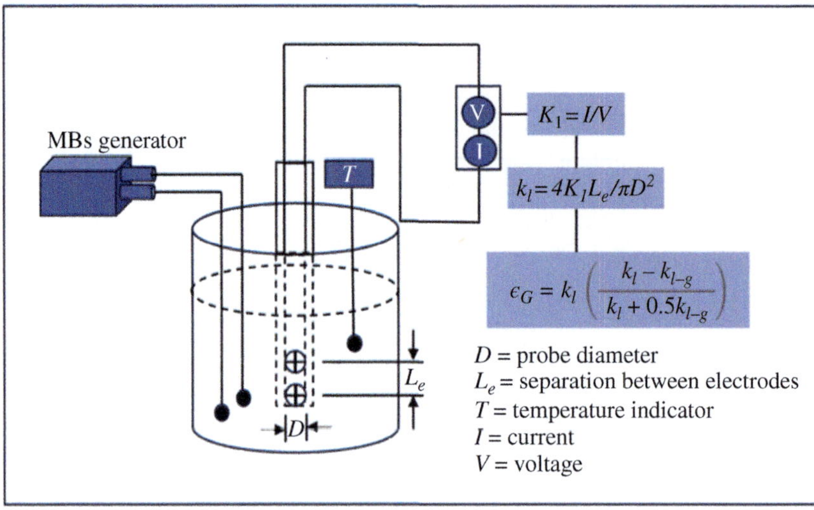

Figure 3.2 The illustration of experimental setup to measure the gas holdup of MBs. Source: Adapted from Parmar and Majumder (2014).

3.2.3.1 Differential Pressure Transmitter

A differential pressure transmitter, an alternative noninvasive method, obtains the average value of the gas holdup at two specified sections of the column. The measurement principle is listed as follows:

$$\epsilon_G = \frac{\Delta P}{(\rho_L - \rho_G)g\Delta H} \approx \frac{\Delta P}{\rho_L g\Delta H} \tag{3.17}$$

The differential pressure transmitter, which, although incapable of directly capturing local gas holdup, offers the most precise and straightforward means of determining the average gas holdup. As a result, it becomes imperative to utilize the ΔP data obtained to rectify the gas holdup measurements obtained via the probe. This method is often used for MB systems for long bubble columns, where a ΔH is at least 1 m, and hence a noticeable ΔP is obtained. For small-height and wider bubble columns or reactors, this method is not suitable for obtaining the ϵ_G.

3.2.3.2 Fiber Optical Probe Method

Another intrusive method for the measurement of gas holdup is the fiber optical probe technique. In this setup, one optical fiber is used as the laser transmitting end, while another acts as the receiving end. During the measurement process, the behavior of light changes based on the phase of the medium being examined. While the fiber tip is inside the gas phase, the laser light is entirely reflected, resulting in a high-voltage signal. But, when the tip is in the liquid phase, the laser light is absorbed by the liquid, resulting in a low-voltage signal. A data acquisition device conducts several photoelectric conversions and amplifies the major signals. Following analog-to-digital conversion, a series of fiber signal differences are obtained after the completion of data collection. The key advantage of this technique is its ability to provide precise and real-time measurement of the bubble hydrodynamic parameters in the gas–liquid bubble column. However, it is important to note that the probe is not capable of detecting smaller bubbles, which means that the recorded values tend to be lower than the actual experimental values. This change in refractive index is the basis for the probe's effectiveness in this application.

The value of gas holdup is determined by averaging the data collected using the fiber optical probe, and subsequently, a correction coefficient is established by comparing it to the value obtained through the differential pressure method. An illustrative schematic diagram depicting the measurement of bubble parameters is presented by Ning et al. (2020). Those measurements are taken at five distinct radial positions, with r_1 representing the center of the circle and r_6 signifying the bed's sidewall radius. It is worth noting that the gas holdup and bubble velocity at the bed's sidewall are both zero. These six radial positions yield valuable information regarding gas holdup and bubble velocity.

The ratio of summation of the average gas holdup for various rings to the cross-sectional area of the bed yields the average gas holdup. The expression for the average gas holdup in integral form is presented as follows:

$$\overline{\epsilon} = \frac{2\pi \int_0^R \epsilon r dr}{\pi R_p^2} \tag{3.18}$$

$$\overline{\epsilon_i} = \frac{\epsilon_{i+1} + \epsilon_i}{2} \tag{3.19}$$

$$S_i = \pi \left(r_{i+1}^2 - r_i^2\right) \tag{3.20}$$

$$\overline{\epsilon} = \frac{\sum_{i=1}^{5} (\epsilon_{i+1} + \epsilon_i) \left(r_{i+1}^2 - r_i^2\right)}{R_p^2} \tag{3.21}$$

3.2.3.3 Electrical Resistance Tomography (ERT) Method

Using the change in the electrical conductivity of the two phases, gas and liquid, or the difference in ERT, the signal of conductivity is converted into gas holdup. This was first proposed by the application of the Maxwell equation.

Subsequently, an image visualization method is employed to enhance the representation of the evolving hydrodynamic traits within the column. The equations involved in Maxwell's transformation are articulated as follows:

$$\bar{\epsilon} = \frac{2k_1 + k_2 - 2k_{mc} - \dfrac{k_{mc}k_2}{k_1}}{k_{mc} - \dfrac{k_2}{k_1}k_{mc} + 2(k_1 - k_2)} \tag{3.22}$$

$$\bar{\epsilon} = \frac{2k_1 - 2k_{mc}}{k_{mc} + 2k_1} \tag{3.23}$$

Within the gas–liquid phase, air is the dispersed phase, resulting in k_2 equal to 0. This relationship is subsequently transformed into Eq. (3.25). To ensure the congruence of the three measured outcomes, three distinct methods of measurement are concurrently applied to gauge the gas holdup under identical operational conditions. The gas holdup measurements obtained through the ERT and differential pressure transmitter are fundamentally consistent. Despite the proximity of both probes used in ERT, there remains a limitation in detecting small or edge bubbles that may pass through the probe. Consequently, the recorded gas holdup value tends to be less compared to the experimental results. The validation of results is illustrated in Figure 3.3.

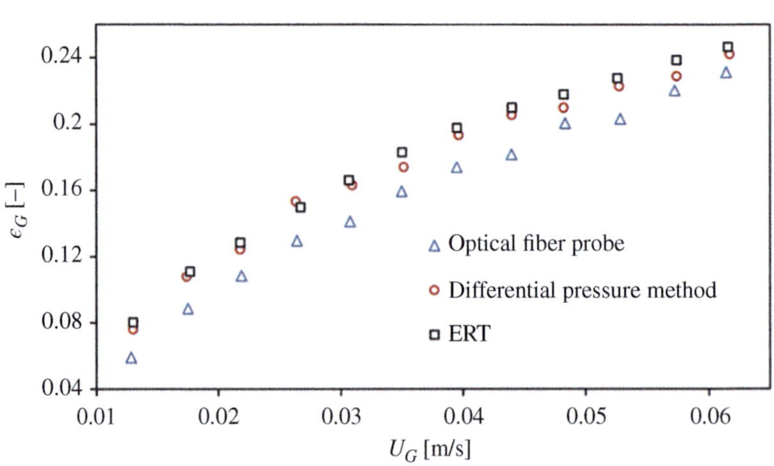

Figure 3.3 The gas holdup, ε_G, *as the function of superficial gas velocities,* measured by ERT, differential pressure method, and optical fiber probe method. Source: Adapted from Ning et al. (2020).

3.2.4 Superficial Gas Velocity

The usual expression for mean bubble rising velocity, u_b, in a gas–liquid bubble column is expressed as the ratio of superficial gas velocity, U_G, and fractional gas holdup ϵ_G (Kawahara et al., 2009),

$$u_b = \frac{U_G}{\epsilon_L} \tag{3.24}$$

whereas U_G in the gas–liquid bubble column is usually defined as:

$$U_G = \frac{V_G}{A_C} \tag{3.25}$$

The various flow regimes in the gas–liquid bubble column can be distinctly categorized into three regimes, such as homogeneous, transition regime, and heterogeneous. These flow regimes are intricately linked to various factors, including column diameter and height, superficial gas velocity, and initial height of the liquid level. Furthermore, the u_b and the type of gas contactor exert significant influence on the observed flow regime. The occurrence of coalescence and bubble breakup is relatively infrequent at lower superficial gas velocities. Also, the design of the gas contactor plays a pivotal role in determining bubble size, subsequently influencing the location of the flow regime transition point. Typically, it is accepted that the generated bubble sizes fall within the range of 2–10 mm when the U_G is below 0.05 m/s. As the U_G escalates, the flow regime undergoes a transition phase. Beyond the point of this transition, where the U_G exceeds a certain threshold, the column experiences heightened turbulent fluctuations, coalescence, bubble breakup, and notable alterations in bubble size. The majority of researchers have focused their studies on investigating the interplay between d_b, U_G, and gas–liquid physical properties, as they significantly impact the hydrodynamic parameters in the bubble column.

In the ***fiber optical probe*** method, U_G of the MBs can be measured based on the time delay between the output signals of two sensors denoted as t_0.

$$U_G = \frac{L_p}{t_0} \tag{3.26}$$

3.3 Correlations of Void Fraction

To predict void fraction in the bubble column, various correlations were proposed for cases involving the use of a typical gas sparger for generation of milli-bubbles. Among these correlations, a majorly accepted correlation was proposed by Akita and Yoshida (1973) involving nonviscous liquids ($\mu_L < 0.02$ Pa s) that satisfies the experimental results.

$$\frac{\epsilon_G}{(1-\epsilon_G)^4} = 0.20\left(\frac{g\rho_L D_c^2}{\sigma}\right)^{1/8}\left(\frac{g\rho_L^2 D_c^3}{\mu_L^2}\right)^{1/12}\left(\frac{U_G}{\sqrt{gD_c}}\right) \tag{3.27}$$

Equation (3.27) was obtained from the experimental data of various gas–liquid systems at $U_G = 0.0053$–0.419 m/s. As a support to this, similar correlations by Hughmark (1967) and Hikita and Kikukawa (1974) were also examined. Hughmark (1967) has derived a correlation considering the liquid properties. Based on his own work and other investigators' experimental data, the ϵ_G could be expressed as:

$$\epsilon_G = \frac{1}{2+\left(\frac{0.35}{U_G}\right)\left(\frac{\rho_L\sigma}{72}\right)^{1/3}} \tag{3.28}$$

Equation (3.28) was obtained from experimental data for air-liquid systems in the range of $u_G = 0.004 - 0.45$ m/s.

3.4 Analysis of Flow Regimes

The scale up of bubble columns is required for the application in larger capacity. While scaling up, it is often required to estimate the flow regime of the bubbles in the column to estimate the total pressure drop. Along with the gas holdup, flow regimes are known to be the important parameter in designing the bubble columns. Two types of flow regimes are observed in the bubble column, such as homogeneous and heterogeneous, where in between a transition occurs. At certain instances, both regimes coexist in the same system, which is dependent on several factors such as type of bubble generator or the gas distributor, bubble column diameter and height, and the physical properties of the phases. If the bubbles are under the homogeneous regime, the bubble will have a uniform spatial distribution for a wide range of U_G, and ε_G. The bubbles are also unaffected by the additional disturbances due to the low gas flow rate. When the bubbles lose their stability at high U_G, with noticeable spatial velocity and ε_G, it is said to have a heterogeneous regime. The usual expression of bubble column dimensions is denoted as the aspect ratio, A = height/diameter, mostly applicable for the heterogeneous regime encountered practically. The visual representation of homogeneous and heterogeneous regimes is shown in Figure 3.4. In the context of MBs, typically two statistical methods are considered to estimate the flow regimes in the MB systems, such as (a) swarm velocity method and (b) drift flux method.

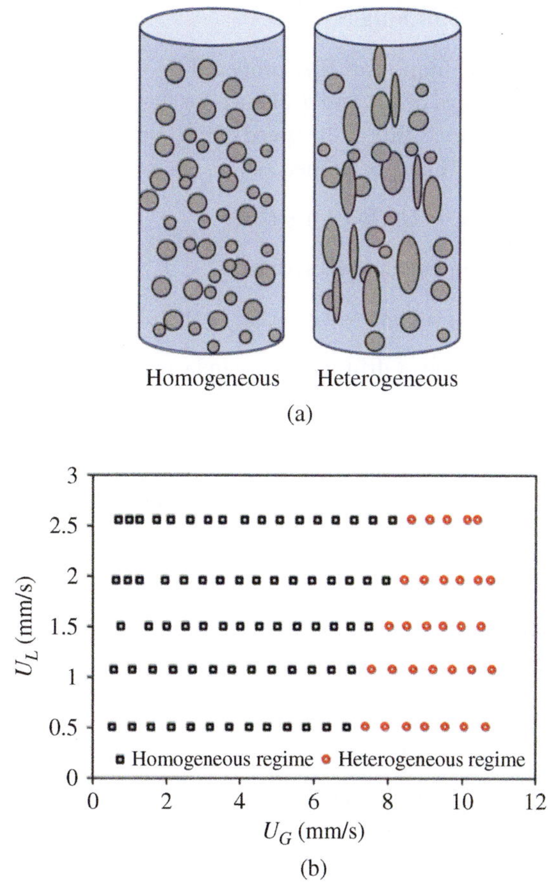

Figure 3.4 (a) Homogeneous flow pattern and heterogeneous flow pattern observed in bubble column and (b) variation of flow regime map with respect to superficial gas velocity in x-axis. Source: Xie et al. (2024)/John Wiley & Sons.

3.4.1 Swarm Velocity Method

Swarm velocity is nothing but the gas phase rise velocity in the bubble column, which occurs in prominent at high gas flow rates. The method was first proposed by Zuber and Findlay (1965) and is widely applicable to the macrobubble columns. The model says that the ratio of superficial gas velocity, U_G, and swarm velocity, U_{swarm}, is a constant for homogeneous flow regime and equivalent to the gas holdup, ε_G.

$$\epsilon_G = \frac{U_G}{U_{swarm}} \tag{3.29}$$

This method is largely useful for macrobubbles, where the value of ϵ_G is high at higher U_G. However, in MB systems, the drift flux method is more acceptable to predict the sensitive ϵ_G and smaller size of MBs.

3.4.2 Drift Flux Method

This technique was first proposed by Wallis (1969) and is applicable for a wide range of U_G and d_b, hence for the MBs. Drift flux velocity, denoted as u_w, is the gas flux at the speed of the two-phase mixture moving through the surface. The u_w is defined as

$$u_w = U_G(1 - \epsilon_G) \pm U_L\epsilon_G \tag{3.30}$$

In a concurrent MB system,

$$u_w = U_G(1 - \epsilon_G) - U_L\epsilon_G \tag{3.31}$$

The value of drift flux (J_E) can be evaluated by considering the U_{swarm} as follows:

$$J_E = U_b(1 - \epsilon_G) \tag{3.32}$$

where U_b is a parameter defined in the drift flux method, as:

$$U_b = U_t\epsilon_G(1 - \epsilon_G)^x \tag{3.33}$$

U_t is the bubble terminal velocity in the column, and x is a fluid-dependent constant, approximately 2 for water. For homogeneous regime, $J_E = u_w$, whereas a transition point arises when $J_E \neq u_w$, and the regime transfers to heterogeneous regime.

3.5 Factors Affecting the Bubble Size and Rise Velocity

The rise of bubbles, or MBs, in low Re liquids is influenced by the physical parameters of the fluid, such as density, viscosity, surface tension, and inertia. When bubbles deviate from the spherical form and exhibit oscillatory motion, surface tension emerges as a predominant factor. In scenarios where the bubble shape remains constant, as observed with MBs, larger bubbles in the same medium tend to rise more rapidly. Additionally, the rise velocity of bubbles is affected by the wall effect, water-phase viscosity, the presence of dispersing agents, and the type of surfactants, all of which contribute to variations in bubble size.

3.5.1 Wall Effect

While the impact of the wall is significantly less in non-Newtonian fluids following the power-law, compared to Newtonian fluids (Chhabra, 2006), the influence becomes minimal when the ratio of d_b to the distance from the column wall at the measurement position is less than 100, as indicated in the study by Krishna et al. (1999). Notably, Haapala et al. (2010) reported a range of values for this ratio, falling between 0.02 and 0.12. Thus, the u_b in the column is a finite number, compared to its rising velocity considered as an infinite volume (u_∞), as defined by Clift et al. (1978) as:

$$\frac{u_b}{u_\infty} = 1 - \frac{9}{16}\frac{d_b}{l_{wall}} \tag{3.34}$$

l_{wall} is the distance of a bubble from a vertical column wall.

3.5.2 Effect of Surfactants

Surfactants have a negative impact on the u_b within a multiphase environment, particularly in the presence of any impurities on the bubble surface (Clift et al., 1978). This phenomenon was elucidated through the Marangoni effect, where changes in the surface tension lead to the development of tangential shear stress, which results in a reduction of the u_b. Frumkin and Levich (1947) initially described this phenomenon, and Cuenot et al. (1997) provided a subsequent review. The value of C_D for contaminated bubbles can be up to seven times higher than for clean bubbles. In addition to that, surface active chemicals or impurities also cause bubble surface stabilization, which prevents the bubble coalescence, hence reducing the surface tension of the bubble.

In practical cases, the bubble surface is often largely contaminated, achieving a no-slip surface. Consequently, it is assumed that contaminated bubbles rise at rates comparable to rigid spheres at low Reynolds numbers, with no apparent bubble surface motion. In that case, the standard drag curve is represented by the correlation proposed by Putnam et al.

$$C_D = \frac{24}{\text{Re}}\left(1 + \frac{1}{6}\text{Re}^{2/3}\right) \text{ for Re} < 1000 \tag{3.35}$$

3.5.3 Effect of Water Viscosity

The presence of surface active groups and pollutants changes the water viscosity, and hence the d_b increases with increasing water viscosity. In the study reported by Kukizaki and Goto (2007), the ratio of the gas (air)/water viscosity on the d_b produced from the Shirasu porous glass membrane was linear with a negative slope when water with 0.3 wt.% sodium dodecyl sulfate solution and different concentrations of polyethylene glycol solution. It is assumed that increasing the viscosity of water slows down the detachment of the MBs, causing more dispersed gaseous phase demand during the formation of MBs. Thus, larger MBs were generated through the membrane pores. However, for other types of microbubble generators (MBGs), the viscosity ratio does not affect the MBs' formation. But later it will contribute to the bubble coalescence while rising along the column.

3.6 Hydrodynamics of Microbubbles Depending on Microbubble Generators

A new gas–liquid contactor or distributor was developed to produce high-density MBs by Yang et al. (2018). It was observed that with high density of MBs, improved the coefficient of mass transfer and enhanced the surface concentration, and hence the overall reaction rate constant was achieved. Bae et al. (2020) generated the MBs under elevated pressure and evaluated the hydrodynamic parameters of MB suspension. Besagni et al. (2018) used a novel needle sparger-type MB contactor in the binary liquid phase consisting of water–moethylene glycol. The method of using different spargers to generate MBs is one of the simplest techniques. It was pragmatic that change in the U_G makes the gas holdup curve concave in shape, which is different from "S-shape" observed in the case of the coarse porous spargers used in conventional bubble columns. Hernadez-Alvarado et al. (2017) experimented on the ε_G, d_b, and interfacial area a, bubble column with MBs in a co-current downflow dispersion, and found that their bubble generation technique can have a higher interfacial area compared to the conventional bubble columns. Another interesting available technique, the foam gun technique, is used as an MB generation, where the ε_G is higher and the bubble density, which is expressed as the foaming effect, is better.

3.6.1 Hydrodynamics of MBs in Foam Gun (Drift Flux Model)

As mentioned earlier, the flow dynamics within an MB column are typically categorized into three regimes: the homogeneous, transition, and heterogeneous regimes. In the context of an MB generator, when the U_G surpasses a certain threshold, the interaction among MBs intensifies, leading to the coalescence of MBs into smaller bubbles. Simultaneously, the rising velocity of bubbles accelerates. With an increase in U_G and ϵ_G, the flow regime changes from the homogeneous and attains a transition regime. Precisely identifying the column flow pattern holds significant importance for its design and practical applications. Researchers commonly employ the $\epsilon_G - U_G$ mapping to roughly determine the flow regime. For this purpose, the drift flux model, recommended by Zuber and Findlay (1965) and expressed in Eq. (3.36), is frequently utilized:

$$\frac{U_G}{\epsilon_G} = C_0(U_G + u_l) + C_1 \tag{3.36}$$

When the bubble column is under the batch operation, the U_G nearly attains 0. So the Eq. (3.36) can be modified to Eq. (3.37).

$$\frac{U_G}{\epsilon_G} = C_0 U_G + C_1 \tag{3.37}$$

C_0 is an indicating the radial distribution-gas holdup and velocity; the value of C_0 approaching 1 signifies the most uniform radial distribution-gas holdup. Consequently, the variation in C_0 becomes instrumental in determining the prevailing flow pattern. As depicted in Figure 3.5, the correlation between drift flux, as defined by Eq. (3.36), U_G is illustrated for both the sintered plate MB distributor and foam gun MB generation system. The sintered plate system resulted in a more pronounced alteration in gas–liquid flow with varying superficial gas velocity. Notably, the results of the two planes were not identical, aligning with the observations in Figure 3.6. Conversely, the flow pattern exhibited greater stability at higher U_G in the foam gun system, particularly providing a uniform bubble regime. This stability is confirmed using the drift flux method, and the alterations in flow patterns at elevated U_G are beneficial for the case of gas–liquid reactions in the MB state. This enhanced stability in the flow pattern at

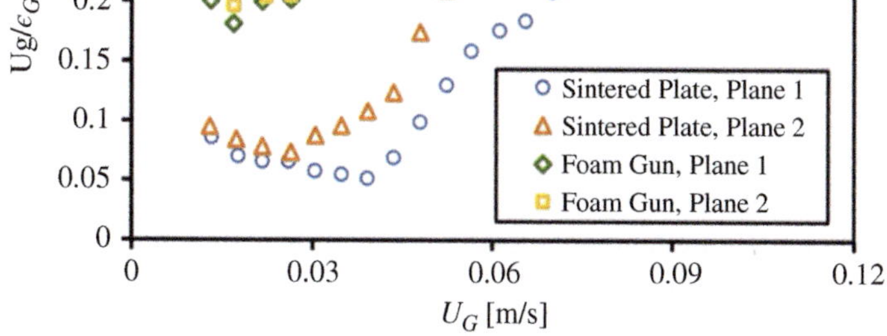

Figure 3.5 Identification of flow pattern using drift flux model, (a) in sintered plate-type MB generator and (b) in foam gun-type MB generator. Source: Ning et al. (2020)/MDPI/CC BY 4.0.

Figure 3.6 The observed gas holdup as a function of superficial gas velocity in two types of MBs generating methods. Source: Reproduced with permission from Ning et al. (2020)/MDPI/CC BY 4.0.

higher superficial gas velocities further supports the facilitation of pertinent gas–liquid reactions within the uniform bubble regime.

In literature, the direct reporting of parameters C_0 and C_1 is often challenging, primarily due to the difficulty in measuring the liquid velocity in two-phase flow. To address this limitation, Wallis (1969) proposed an alternative drift flux theory for assessing the flow pattern, modifying the original drift flux model as expressed in Eq. (3.38):

$$u_{df} = U_G(1 - \epsilon_G)^{n-1} \pm u_l \epsilon_G \tag{3.38}$$

Here, u_{df} represents the drift flux velocity, and n is the Richardson–Zaki index for low viscosity and high Re systems, typically assigned to a value of 2.39. In batch mode, u_l is 0, and with (n–1) close to 1, which is the case similar to low-viscosity fluids, Eq. (3.38) simplifies to Eq. (3.39):

$$u_{df} = U_G(1 - \epsilon_G) \tag{3.39}$$

As the drift flux theory remains stable only for the uniform regime, the change in slope of the critical point becomes the transition point for the flow regime. Consequently, drift flux theory is proven effective in determining the transition point to obtain the gas holdup with varying flow patterns. This information is particularly crucial for understanding sensitive gas–liquid reactions. Comparing the findings with Figure 3.6, the drift flux velocity profile concerning gas holdup closely mirrors the profile depicting ϵ_G with U_G. Additionally, the comparative study of drift flux velocity, u_{df} with ϵ_G is more accurate than u_{df}/ϵ_G and ϵ_G. This enhanced accuracy holds significant importance in comprehending the intricate exchange of gas and liquid phases in the system.

3.6.2 Hydrodynamics of Microbubbles in Sintered Plate Microbubble Generators

A case of the hydrodynamics of MBs generated by sintered plate MBGs is proposed by Kukizaki and Goto (2007). A linear relation between the d_b and the viscosity ratio was expressed by the equation:

$$\log(d_b) = -0.304 \log \eta_r + 0.98 \tag{3.40}$$

where η_r is the viscosity ratio of the gas and water phase. During the generation of bubbles, the air (dispersed phase) flux through the membrane pores, where the flux, $J_{d,}$ is expressed following the Hagen–Poiseuille law:

$$J_D = \frac{k(\Delta P - P_c)}{\eta_d R_m} = \frac{k}{\eta_d} \frac{\varepsilon_m (D_m)^2 (\Delta P - P_c)}{32\tau^2 \Delta x} \tag{3.41}$$

where k is a constant, proportional to active pores, which is equivalent to the fraction of pores that take part in the formation of bubbles, and is almost a constant for a relatively low transmembrane/capillary pressure range. This correlation shows that the J_d decreases with increasing η_d at a constant transmembrane pressure, which is higher than the capillary pressure. The expression of capillary pressure is given by:

$$P_c = \frac{4\gamma_r \cos\theta}{D_m} \tag{3.42}$$

where γ_r is the relative surface tension between the gas and water phase. Typically, the bubble/droplet formation begins at the surface of the membrane when the transmembrane pressure is more than the capillary pressure. Typically, the bubbles/droplets are generated at a transmembrane/capillary pressure ratio, $\Delta P/Pc = 1.1$.

In bubble formation from porous medium, another important factor is the time required for the generation of a bubble from the active pore, denoted as t_f. The bubble formation time is calculated from the ratio of V_{dr} and V_G through a single pore:

$$t_f = \frac{V_{dr}}{V_G} = \frac{\frac{\pi}{6} d_b^3}{\frac{\pi}{4} d_b^2 v} \tag{3.43}$$

where V_{dr} is the volume of a bubble and v_d is the velocity of dispersed phase through a single pore, which is expressed as follows:

$$v_d = \frac{J_D}{\varepsilon_m k} \tag{3.44}$$

By rearranging Eq. (3.41) to Eq. (3.42), bubble diameter is given by:

$$d_b = \left\{ \frac{3}{2(\Delta P - P_c)\varepsilon_m R_m} \frac{t_f}{\eta_d} \right\}^{1/3} \tag{3.45}$$

At constant viscosity of the water phase, t_f independent of the dispersed-phase viscosity and only depends on the transmembrane pressure and the dispersing agent. In Eq. (3.45), dispersed-phase viscosity does not affect d_b, ΔP, and Pc, and the value obtained for d_b decreases with rising dispersed-phase viscosity.

3.7 Hydrodynamics of Microbubbles Flow in Pipes

MB suspensions exhibit non-Newtonian shear-thinning (pseudoplastic) behavior. The fluid dynamics of an MB system through a pipe can be elucidated using the internal flow model. The rate of energy dissipation is incorporated while modeling the dynamic phase interaction. The pressure drop is calculated by the mechanical energy balance, representing either to surpass the frictional forces and/or the energy dissipation. The following assumptions are considered for the model operation:

(i) Isothermal steady-state flow is attained with uniform ε_G.

(ii) The effect of acceleration is negligible, considering no interphase mass transfer.

(iii) Uniform frictional loss is assumed across the entire pipe.

Experimentally, the total ΔP is derived as the frictional ΔP attributable to the liquid flow, neglecting any hydrostatic pressure drop in vertical direction. This is be expressed as:

$$\Delta P_{l-MB} = \left(\frac{1}{V_m A_p} \right) \left(\frac{3}{4} C_D \rho_L U_L^3 A_p \frac{1-\epsilon_L}{\epsilon_L^3} + \frac{L_t}{d_b} + \frac{\pi d_p U_L \sigma_{lw}}{\epsilon_L} + \frac{U_L \sigma_{l-mb}}{a\epsilon_L} \right) \tag{3.46}$$

where ΔP_{l-MB} represents the frictional ΔP in the liquid-MB suspension. The hydrodynamic drag coefficient in pipe flow is given by the equation:

$$C_m = \frac{4}{3} \left(\frac{\sigma_1 \Delta P_{flo} V_m A_p}{\epsilon_l^3} - \frac{\pi d_p U_L \sigma_{lw}}{\epsilon_l} - \frac{U_L \sigma_{l-mb}}{a\epsilon_l} \right) \left(\frac{\epsilon_l^3 d_b}{(1-\epsilon_1)\rho_l U_L^3 A_p L_t} \right) \tag{3.47}$$

Here, σ_I is the force required for stretching the interface by a unit distance. The liquid and the solid wall surface tension (σ_{lw}) is determined using Young's equation:

$$\sigma_1 cos\theta_1 = \sigma_{lw} \tag{3.48}$$

ΔP_{flo} is the frictional pressure drop of the liquid (in the absence of bubbles) and is calculated by using Fanning's equation:

$$\Delta P_{flo} = \frac{2f_{lo}\rho_l U_L^2 L_t}{d_p} \tag{3.49}$$

f_{lo} is the friction factor, which should be taken appropriately for laminar and turbulent flow.

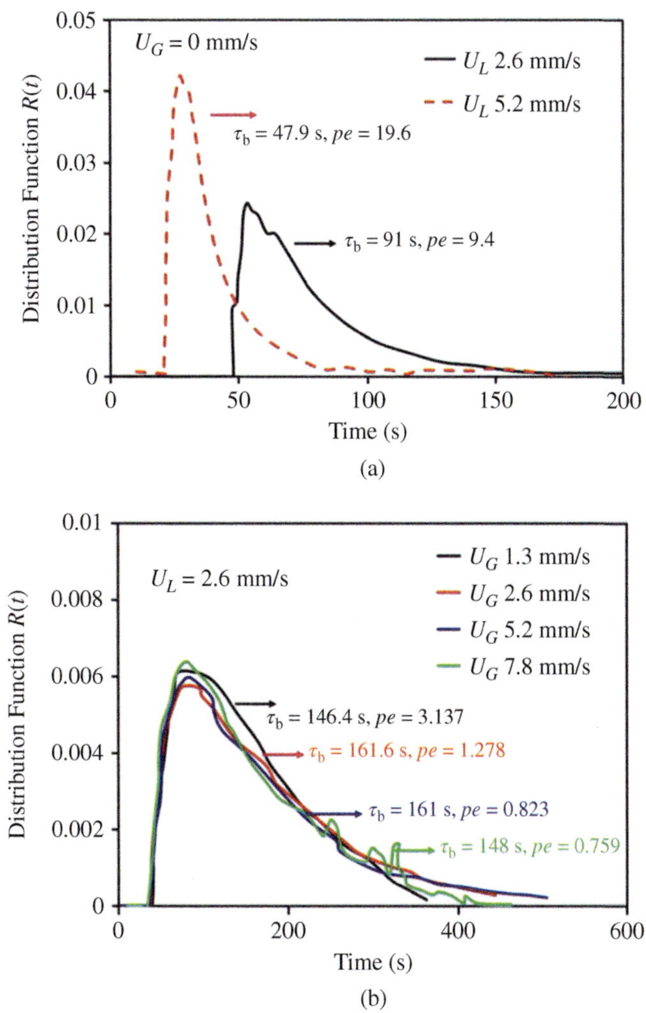

Figure 3.7 The values of RTDs in MB column for (a) $U_G = 0$ mm/s and (b) $U_G > 0$ mm/s. Source: Xie et al. (2024)/John Wiley & Sons.

3.8 The Residence Time of Microbubbles in the Column

Due to the slow and continuous rise of MBs in the liquid phase, the residence time of the MBs is known to be longer than the usual macrobubbles. The MBs' residence time in the column is calculated by the residence-time-distribution (RTD) method. From the RTD, further, the mean residence time is calculated using the following expressions.

$$R(t) = \frac{C(t)}{\int_{o}^{\infty} C(t)dt} \tag{3.50}$$

$$\tau_b = \int_{0}^{\infty} tR(t)dt \tag{3.51}$$

where $R(t)$ is the RTD function, $C(t)$ is the concentration of tracer measured with respect to time, and τ_b is the bubble residence time. $C(t)$ is measured by the spectrometer. An RTD curve for MB is shown in Figure 3.7 at different U_G values.

Nomenclature

Notation		SI unit
a	interfacial area per unit volume	m^2/m^3
A_e	electrode area	m^2
A_c	cross-sectional area of column	m^2
A_p	cross-sectional area of pipe	m^3
C_0	radial distribution	unit less
C_D	drag coefficient of the microbubble mixture	unit less
C_m	drag coefficient of microbubble mixture in a pipe flow	unit less
$C(t)$	concentration of tracer measured with respect to time	mol/L
d_b	bubble diameter	m
D_c	column diameter	m
Dm	mean pore diameter	m
f_{lo}	friction factor	unit less
g	gravitational constant	$9.8\ m/s^2$
ΔH	liquid height difference in column	m
I	current employed on cell	Amp
J_D	dispersed-phase flux	$m^3/m^2.s$
J_E	drift flux	$m^3/m^2.s$

Notation		SI unit
k	proportion constant of active pores	unit less
k_1	conductivity of continuous phase (liquid)	m.s/cm
k_2	conductivity of disperse phase (air)	m.s/cm
k_l	liquid electrical conductivity	s/m
K_l	conductance of liquid	s
$k_{l\text{-}d}$	electrical conductivities of the liquid–gas mixture	S/m
k_{mc}	conductivity measured during experiments	m.s/cm
l_{wall}	distance of bubble from vertical column wall	m
L	chamber length	m
L_e	separation between the electrodes	m
L_p	l is the distance between the two probes	m
Lt	length of the test section	m
n	Richardson–Zaki index	unit less
P	partial pressure	Pa
P_c	capillary pressure	Pa
P_G	gas pressure	Pa
P_L	liquid pressure	Pa
ΔP	pressure drop	Pa
ΔP_{flo}	frictional pressure drop of liquid without bubble	Pa
ΔP_{l-MB}	frictional ΔP in the liquid-MB suspension	Pa
r	radius of ring (Ning et al., 2020)	m
Re_b	bubble Reynolds number	
R_m	hydrodynamic resistance of the membrane	1/m
R_p	diameter of the measuring plane	m
$R(t)$	mean residence time	s
t	time	s
t_0	time for the bubble to pass through the probe	s
t_f	bubble/droplet formation time	s
$\vec{u}_b$	bubble rising velocity vector	m/s
$\vec{u}_l$	liquid velocity vector	m/s
u_b	bubble rising velocity	m/s
U_b	parameter in drift flux	$kg/m^2.s$
$u_{b(H\text{-}R)}$	bubble rising velocity as per Hadamard–Rybczynski equation	m/s
u_{df}	drift flux velocity	m/s
u_∞	rising velocity considering infinite volume	m/s
u_l	liquid velocity	m/s

(Continued)

Notation		SI unit
u_w	gas flux at the speed of the two-phase mixture	$kg/m^2.s$
U_G	superficial gas velocity	m/s
U_L	superficial Liquid velocity	m/s
U_{swarm}	swarm velocity	m/s
U_t	bubble terminal velocity	m/s
v_d	velocity of dispersed phase through a single pore	m/s
V_{dr}	volume of a bubble/droplet	m^3
V_G	gas flow rate	m^3/s
V_m	velocity of the water and MB mixture	m/s
V_t	voltage	V
Δx	membrane thickness	m

Greek Nomenclature

Notation		SI unit
γ_r	relative surface tension between the gas and water phase	unit less
ε	average gas holdup for various rings	unit less
ε_d	volume fraction of a dispersed nonconductive phase	unit less
ε_G	gas void fraction or holdup	unit less
ε_L	liquid holdup	unit less
$\overline{\varepsilon}$	integral of average gas holdup for various rings	unit less
ε_m	membrane porosity	unit less
ηc	viscosity of continuous phase	Pa.s
ηd	viscosity of dispersed phase	Pa.s
η_r	viscosity ratio of the dispersed phase to the continuous phase	Pa.s
θ	contact angle between the membrane surface and liquid	radian
μ_G	viscosity of gas	Pa.s
μ_G	viscosity of gas	Pa.s
μ_L	viscosity of liquid (water)	Pa.s
ρ_G	density of gas	kg/m^3
ρ_L	density of water	kg/m^3
ρm	density of gas–water mixture	kg/m^3
v_L	kinematic viscosity of liquid	m^2/s
σ	surface tension of the liquid	N/m

Notation		SI unit
σ_I	force required for stretching the interface by a unit distance	N/m
σ_{l-mb}	surface tension between liquid and microbubble	N/m
σ_{lw}	surface tension at the boundary between the liquid and the pipe wall	N/m
τ	pore tortuosity	unit less
τ_b	bubble residence time	s

References

Akita, K. and Yoshida, F. (1973). Gas holdup and volumetric mass transfer coefficient in bubble columns. Ind. Eng. Chem. Proc. Des. Dev. 12: 76–80.

Bae, K., Go, G., Noh, N. et al. (2020). Bubble characteristics in pressurized bubble column associated with micro-bubble dispersion. Chem. Eng. J. 386: 121339.

Besagni, G., Gallzaaini, L., and Inzoli, F. (2018). Effect of gas sparger design on bubble column hydrodynamics using pure and binary liquid phases. Chem. Eng. Sci. 176: 116–126.

Chhabra, R.P. (2006). *Bubbles, Drops, and Particles in Non-Newtonian Fluids*, 2nde. Boca Raton: CRC Press.

Clift, R., Grace, J.R., and Weber, M.E. (1978). *Slow Viscous Flow Past Spheres in Bubbles, Drops and Particles*. Academic Press.

Cuenot, B., Magnaudet, J., and Spennato, B. (1997). The effects of slightly soluble surfactants on the flow around a spherical bubble. J. Fluid Mech. 339: 25–53.

Frumkin, A.N. and Levich, V.G. (1947). On surfactant sand interfacial motion. Zh. Fiz. Khim. 21: 1183–1204.

Haapalaa, A., Honkanenb, M., Liimatainena, H. et al. (2010). Hydrodynamic drag and rise velocity of microbubbles in papermaking process waters. Chem. Eng. J. 162: 956–964.

Hadamard, J. (1911). Mouvement permanent lent d'une sphère liquide et visqueuse dans un liquide visqueux. C. R. Math. Acad. Sci. 152: 1735–1738.

Hernandez-Alvarado, F., Kalaga, D., Turney, D. et al. (2017). Void fraction, bubble size and interfacial area measurements in co-current downflow bubble column reactor with microbubble dispersion. Chem. Eng. Sci. 168: 403–413.

Hikita, H. and Kikukawa, H. (1974). Liquid-phase mixing in bubble columns: effect of liquid properties. Chem. Eng. J. 8: 191–197.

Hughmark, G.A. (1967). Holdup and mass transfer in bubble columns. Ind. Eng. Chem. Proc. Des. Dev. 6: 218–220.

Karamah, E.F., Bismo, S., Annasari, L., and Purwanto, W.W. (2010). Mass transfer study on micro-bubbles ozonation in a bubble column. Int. J. Chem. Eng. Res. 2: 243–252.

Kasper, C. (1940). The theory of the potential and the technical practice of electrodeposition. Trans. Electrochem. Soc. 77: 353–384.

Kawahara, A., Sadatomi, M., Matsuyama, F., and Matsuura, H. (2009). Prediction of micro-bubble dissolution characteristics in water and seawater. Exp. Therm. Fluid. Sci. 33: 883–894.

Khuntia, S., Majumder, S.K., and Ghosh, P. (2013). Removal of ammonia from water by ozone microbubbles. Ind. Eng. Chem. Res. 52: 318–326.

Krishna, R., Urseanu, M.I., Baten, J.M.V., and Ellenberger, J. (1999). Wall effects on the rise of single gas bubbles in liquids. Int. Commun. Heat Mass Transfer 26: 781–790.

Kukizaki, M. and Goto, M. (2007). Spontaneous formation behavior of uniform-sized microbubbles from Shirasu porous glass (SPG) membranes in the absence of water-phase flow. Colloids Surf. A: Physicochem. Eng. Aspects 296: 174–181.

Maxwell, J.C. (1892). *A Treatise on Electricity and Magnetism*, 3rde. London: Oxford University Press.

Mei, R., Klaussner, J.F., and Lawrence, C.J. (1994). A note on the history force on a spherical bubble at finite Reynolds number. Phys. Fluids 6: 418–420.

Moore, D.W. (1963). The boundary layer on a spherical gas bubble. J. Fluid Mech. 16: 161–176.

Ning, S., He, G., Ma, L. et al. (2020). Effects of the microbubble generation mode on hydrodynamic parameters in gas–liquid bubble columns. Processes 663: 1–13.

Parmar, R. and Majumder, S.K. (2014). Hydrodynamics of microbubble suspension flow in pipes. Ind. Eng. Chem. Res. 53: 3689–3701.

Shen, Y., Longo, M.L., and Powell, R.L. (2008). Stability and rheological behavior of concentrated monodisperse food emulsifier coated microbubble suspensions. J. Colloid Interface Sci. 327: 204–210.

Takahashi, M. (2005). ζ Potential of microbubbles in aqueous solutions: electrical properties of the gas–water interface. J. Phys. Chem. B 109: 21858–21864.

Wallis, G.B. (1969). *One Dimensional Two-Phase Flow*. New York: McGraw-Hill.

Xie, B., Liu, W., Jiang, Y. et al. (2024). Hydrodynamics and model of gas-liquid flow in microbubble column reactors. AIChE J. 70: 18273.

Yang, Z., Liu, M., and Wang, X. (2018). Experiment study and modelling of novel mini-bubble column photocatalytic reactor with multiple micro-bubbles. Chem. Eng. Process. 124: 269–281.

Zuber, N. and Findlay, J. (1965). Average volumetric concentration in two-phase flow systems. J. Heat Trans. 87: 453.

4

Radical Generation by Microbubbles

4.1 Introduction

Microbubbles (MBs) while rising inside water undergo a fascinating underwater contraction because of the rapid dissipation of internal gas. This captivating phenomenon is ascribed to the combination of an expansive surface area-to-volume ratio and the gas-pressurized state induced by surface tension, as outlined by Tsuge (2014). The extended residence time of MBs in water leads to their gradual reduction in size and eventual collapse below the water's surface, preventing their rise to the upper layer. Furthermore, scientific inquiry has unveiled a noteworthy aspect of these collapsing MBs: the formation of hydroxyl radicals, $OH^{\bullet}$, generating even under strongly acidic conditions, all without the need for dynamic stimuli such as photoactivation, ultrasound, or substantial pressure differentials. This distinctive property imparts remarkable versatility to these tiny bubbles, rendering their remarkable practical applications, including the area of wastewater treatment and medical applications.

It is essential to highlight that the process of radical generation from MBs differs from that observed in acoustic cavitation, a phenomenon characterized by the creation of extremely high temperatures resulting from the adiabatic compression of cavitation bubbles. Notably, the collapse rate of MBs is notably slower in comparison to cavitation bubbles induced by ultrasound. Although MBs lack the capability to decompose perfluorooctanoic acid, sonochemical action achieves decomposition through direct pyrolysis reactions, as demonstrated by Moriwaki et al. (2005). The collapse of MBs unfolds over tens of seconds, a stark contrast to the microseconds it takes for cavitation bubbles induced by ultrasound to collapse. Given the relatively sluggish collapse speed of MBs, it appears unlikely that adiabatic compression plays a role in generating the exceedingly high temperatures required for free radical generation, as indicated by Yasui et al. (2018).

Microbubbles: Engineering Aspects and Industrial Applications, First Edition. Snigdha Khuntia.

Existing literature suggests that the mechanism behind radical generation is intricately tied to the electrical properties exhibited during the collapse of MBs. This chapter will delve into existing studies on the free radical generation capability of MBs, their mechanism, quantification studies, and the influence of various parameters.

4.2 Radical Generation in O_2, N_2, and Air Microbubbles

The surface charge and free radical generation exhibited by MBs is a topic of particular fascination. It is a well-known fact that when a bubble is immersed in the aqueous phase, adsorption of ions at the gas–liquid interface makes the bubble surface charged, as elucidated by Takahashi (2005). The existence of a charged surface on the bubble significantly influences the surface distribution of ions, resulting in the formation of an electric double layer at the interface. The strength of this double layer is commonly assessed using the zeta (ζ) potential. Pioneering work by Takahashi et al. (2007a) first documented the intriguing finding that, even in the absence of dynamic stimuli, the generation of free radicals can be achieved through the collapse of MBs. Their research not only highlighted this phenomenon but also underscored the utility of MBs in influencing the formation of gas hydrates. This influence stems from their capacity to modify nucleation conditions and exhibit efficient solubility of gases. The intricate correlation among the internal gas pressure as well as the MB diameter, d_b, finds expression through the Young–Laplace eq.

$$P = P_L + \frac{4\sigma}{d_b} \tag{4.1}$$

In accordance with Henry's law, the quantity of dissolved gas in a contracting bubble escalates proportionally with the rising gas pressure. As illustrated in Figure 4.1, the region encompassing a MB undergoes an apparent transformation in the pressure–temperature (P–T) diagram, favoring hydrate nucleation. This distinctive trait stands out as a hallmark of MBs. Furthermore, research has unveiled that collapsing MBs give rise to free radicals, presenting a valuable attribute for various technical applications, notably in the decomposition of organic chemicals.

In their study, Li et al. (2009) selected phenol as a representative pollutant to investigate the variance in oxidation efficiency among air, oxygen, and nitrogen MBs across different pH levels. Notably, the findings revealed that oxygen MBs exhibited heightened efficiency under acidic conditions compared to both air and nitrogen MBs in the oxidation of organic compounds.

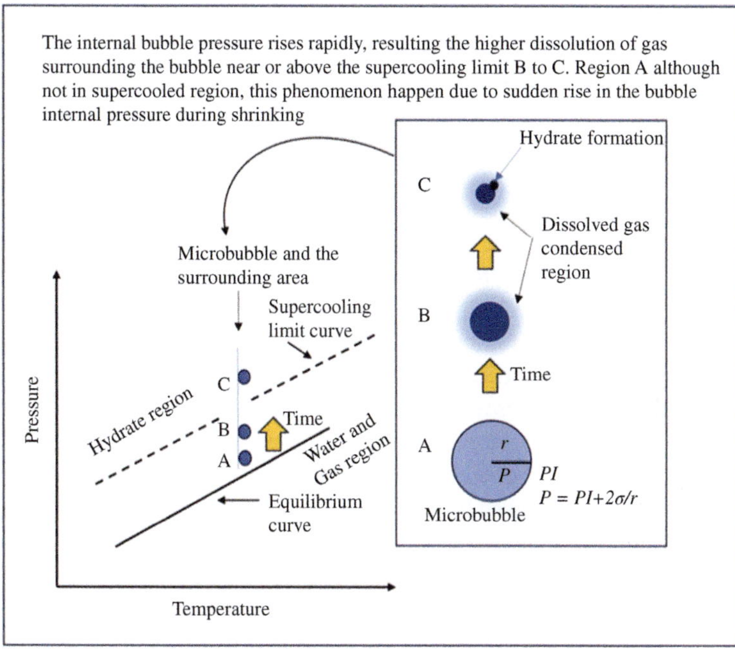

Figure 4.1 The shrinkage of MBs and the corresponding *P*–*T* diagram proposed by Takahashi et al. during the nucleation process. Source: Takahashi et al. (2003)/American Chemical Society.

The qualitative identification of the generated free radicals was conducted through the electron spin resonance (ESR) spin-trap method. Despite the qualitative detection, the study did not provide quantification of the radicals generated by the MBs.

4.3 Radical Generation in O_3 Microbubbles

The utilization of O_3 MBs holds significant importance in contemporary times due to the potent oxidizing capabilities of O_3, enabling the degradation of numerous organic and some inorganic pollutants in water. While ozonation is recognized as a valuable process, it comes with the drawback of being costly due to the expense of generation and the gas's limited solubility in water. The water solubility of O_3 is relatively low, ranging from 1.0×10^{-6} to 1.3×10^{-4} mol/m^3·Pa. Although notably higher than oxygen and nitrogen gases, it still poses challenges as O_3 gas transferred into water can remain in the vapor phase without effectively reacting with the micropollutants in water. Moreover, an excess of ozone released into the atmosphere is considered a health hazard. To overcome these challenges, various pressurized ozone contactors have been employed to maximize ozone solubility in water. A promising solution to enhance solubility involves the use of O_3 MBs (O_3-MBs). The internal pressure of the MBs aids in the dissolution of gaseous ozone in water, extending its residence time. However, it is important to note that under alkaline conditions, the unstable O_3 can release $OH^•$ radicals in the aqueous medium. This aspect has been explored by many researchers as a potential avenue for generating potent $OH^•$ radicals.

A notable contributor to the field of MB technology and radical generation is Takahashi and his team (Takahashi et al., 2012), who have delved into the removal of photoresists using O_3-MBs. Their research has revealed that O_3-MBs can oxidize organic compounds in two pathways: direct molecular O_3 reaction with the compound and the generation of $OH^•$ radicals through the collapse of MBs followed by their reaction with the compound (considered as the indirect approach of reaction). The MBs also demonstrated an enhancement in ozone dissolution ranging from 25% to 70% at various input O_3 concentrations (Figure 4.2 (a) and (b)). This innovative approach showcases the multifaceted benefits of O_3-MBs in water treatment processes.

A well-established explanation for producing $OH^•$ radicals from gas MBs involves the collapse process. During this process, the accelerated shrinking of the interface causes the accumulation of ions near the interface, leading to a swift rise in ζ potential. The profound environmental shift resulting from the gas-water interface extinction initiates $OH^•$ radical generation through the dispersion of higher chemical potential gathered at the interface. In the case of O_3-MBs, the concentration of $OH^•$ radicals generated is largely dependent on pH, a topic that will be explored later in this chapter.

4.4 Application of Catalysts for Radical Generation

In order to stand out among existing technologies, it is imperative to both accelerate and quantify the radicals produced through MB collapse. The quantity of $OH^•$ radicals generated due to the collapse of air MBs has notably fallen short for practical applications till now. Advanced oxidation processes (AOPs) commonly employ catalysts like Fe and H_2O_2 for free radical generation. Similarly, in water treatment, solid powder catalysts such as Fe, activated carbon, Co, and others are utilized to generate free

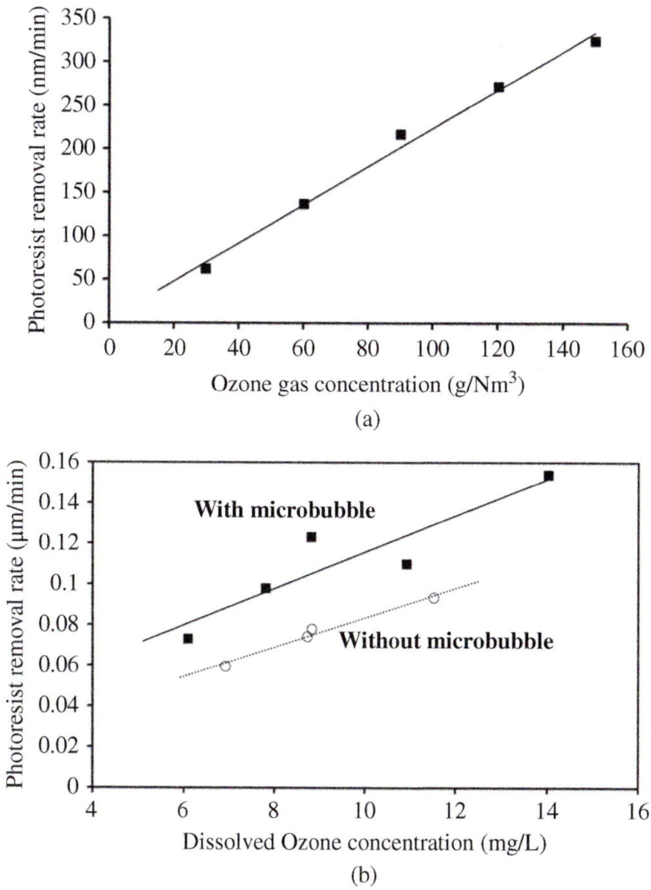

Figure 4.2 (a) Effect of concentration of O_3 and (b) effect of MBs on the rate of photoresist removal. Source: Takahashi et al. (2012)/American Chemical Society.

radicals. The potential of MBs to generate free radicals with the assistance of these catalysts has been thoroughly examined by researchers over the past decades.

Numerous organic compounds, such as polyvinyl alcohol (PVA) and dimethyl sulfoxide (DMSO), exhibit low reactivity toward O_3 (PVA and DMSO are also used as the radical probe). Additionally, some pollutants and intermediates prove challenging to oxidize through O_3 alone. Therefore, the abundance of $OH^•$ radicals generated remains a focal point in research, emerging as a priority area for addressing the limitations and expanding the applicability of MB technology.

4.4.1 Homogeneous Catalytic Process

In the context of homogeneous catalytic processes, hydrogen peroxide (H_2O_2) stands out as a prevalent choice owing to its robust oxidizing capabilities (Khuntia et al., 2021; Khuntia and Mohan, 2022; Dalal et al., 2023). The reaction scheme involves the combination of H_2O_2 and O_3, resulting in the formation of $OH^{\bullet}$ radicals. The sequential steps to produce $OH^{\bullet}$ radicals in this reaction are as follows:

$$O_3 + OH^- \rightarrow HO_2^- + O_2 \qquad k = 40\ \text{L/mol.s} \tag{4.2}$$

$$O_3 + HO_2^- \rightarrow {}^{\bullet}O_3^- + HO_2^{\bullet} \qquad k = 2.2 \times 106\ \text{L/mol.s} \tag{4.3}$$

$$HO_2^{\bullet} + OH^- \rightleftharpoons {}^{\bullet}O_2^- + H_2O \qquad pK = 4.8 \tag{4.4}$$

$$O_2^{-\bullet} + O_3 \rightarrow O_3^{-\bullet} + O_2 \qquad k = 1.6 \times 109\ \text{L/mol.s} \tag{4.5}$$

$$O_3^{-\bullet} + H_2O \rightarrow OH^{\bullet} + O_2 + OH^- \qquad k = 20 - 30\ \text{1/s} \tag{4.6}$$

$$O_3^{-\bullet} + OH^{\bullet} \rightarrow O_2^{-\bullet} + HO_2^{\bullet} \qquad k = 6 \times 109\ \text{L/mol.s} \tag{4.7}$$

$$O_3^{-\bullet} + OH^{\bullet} \rightarrow O_3 + OH^- \qquad k = 2.5 \times 109\ \text{L/mol.s} \tag{4.8}$$

$$OH^{\bullet} + O_3 \rightarrow HO_2^{\bullet} + O_2 \qquad k = 3 \times 109\ \text{L/mol.s} \tag{4.9}$$

$O_3 + UV + MB$ is another method of generating more $OH^{\bullet}$ radicals in water (Glaze et al., 1987):

$$O_3 + h\nu(\lambda < 310\ \text{nm}) \rightarrow O_2 + O(^1D) \tag{4.10}$$

$$O(^1D) + H_2O \rightarrow 2OH^{\bullet} \tag{4.11}$$

Earlier assumptions, as proposed by Prengle et al. (1975), suggested that the photolysis of ozone in water would follow a comparable pathway. In this process, hydrogen peroxide is generated, and any $OH^{\bullet}$ radicals, if formed at all, are constrained within the solvent cage, without escaping. The overarching reaction is as follows:

$$(4.1)\ O_3 + h\nu + H_2O \longrightarrow H_2O_2 \tag{4.12}$$

Equation (4.12) might give the impression that the O_3/UV and $O_3 + H_2O_2$ processes are equivalent: in the former, hydrogen peroxide is essentially formed in situ, as opposed to being externally introduced (Khuntia et al., 2020a, 2020b).

Research into the $O_3 + H_2O_2 + MB$ system is prevalent, whereas the exploration of $O_3 + UV + MB$ is less frequent. The primary deterrent is the elevated energy consumption associated with the combination of the MB system and UV irradiation. Azuma et al. (2019) conducted an extensive comparison of the homogeneous catalytic MBs system for 39 pharmaceutical compounds, and a selection of the findings is presented in Table 4.1.

The addition of UV and H_2O_2 with O_3-MB demonstrated a noteworthy enhancement in the overall reaction rate constant, raging 2.9–5.5 times higher compared to only O_3 and O_3-MB systems. Moreover, the treatment with MBs led to a substantial increase in overall O_3 utilization, reaching up to 2.8 times better than O_3 treatment alone. These findings suggest that the integrated application of O_3-MB with additional UV and/or H_2O_2 holds promise for mitigating the environmental impact of pharmaceutical pollutants discharged into rivers. However, caution is advised when considering $O_3 + UV + H_2O_2 + MB$, as this process could pose challenges due to its potentially high energy consumption.

In addition to incorporating H_2O_2 and UV, numerous transition metal catalysts, classified as homogeneous catalysts, are extensively utilized. The application of metal ion catalysts such as Fe(II) or Fe(III), Co(II), Ni(II), Cu(II), Zn(II), Mn(II), Ag(II), and Ti(II) is notably prevalent in the decolorization of various dyes (Rodriguez-lopez et al., 2013; Wu et al., 2008). These metal catalysts play a crucial role in augmenting hydroxyl radical generation, facilitating compound oxidation through both direct and indirect pathways (Pi et al., 2003; Rivas et al., 2001). The inclusion of these metals also enhances total organic carbon (TOC) removal efficiency. Many mechanistic studies support the findings that the generation of $OH^\bullet$ radicals, which emerge from the reaction of O_3 and metal ions, significantly amplifies the oxidation activity against pollutants. According to reports by Wu et al. (2008), the interaction of Fe(II) and Fe(III) with ozone results in the formation of $OH^\bullet$ radicals, as outlined below:

$$Fe^{2+} + O_3 \rightarrow Fe^{3+} + O_3^\bullet \tag{4.13}$$

$$O_3^\bullet + H^+ \rightarrow O_2 + OH^\bullet \tag{4.14}$$

$$Fe^{2+} + O_3 \rightarrow FeO^{2+} + O_2 \tag{4.15}$$

$$FeO^{2+} + H_2O \rightarrow Fe^{3+} + OH^\bullet + OH^- \tag{4.16}$$

$$Fe^{3+} + O_3 + H_2O \rightarrow FeO^{2+} + H^+ + OH^\bullet + O_2 \tag{4.17}$$

$$2HO_2^\bullet \rightarrow H_2O_2 + O_2 \tag{4.18}$$

Table 4.1 Pharmaceutical compounds degradation using various catalytic O_3 + MB processes.

	O_3-based system (% removal of the compound)				O3 + MB-based system (% removal of the compound)			
Compound type	O_3	O_3/UV	O_3/H_2O_2	O_3/UV/H_2O_2	O_3	O_3/UV	O_3/H_2O_2	O_3/UV/H_2O_2
Antiviral (Famciclovir)	63.3	31.0	22.1	19.2	66.1	65.8	78.9	96.0
Antibacterial (Ciprofloxicin)	100.0	76.7	98.4	48.4	100.0	99.0	90.1	98.5
Antibacterial (Clarithromycin)	94.8	79.6	45.5	60.2	100.0	96.5	87.0	99.4
Anticancer (Bicalutamide)	42.9	47.6	39.2	15.4	38.7	35.3	10.3	98.9
Anticancer Cyclophos-phamide	21.5	0.0	13.9	12.0	14.0	30.2	26.1	32.0
Psychotropic (Sulpiride)	85.4	32.8	67.4	39.7	98.6	80.3	92.7	82.1

$$FeO^{2+} + H_2O_2 \rightarrow Fe^{3+} + OH^{\bullet} + OH^{-} \quad (4.19)$$

$$Fe^{3+} + H_2O_2 \rightarrow Fe^{2+} + HO_2^{\bullet} + H^{+} \quad (4.20)$$

It has been stated that the initiation of O_3 dissociation occurs in alkaline pH through OH^-, while in acidic pH, thermal dissociation of molecular O_3 produces some radicals like $OH^{\bullet}$. These intermediates spontaneously recombine, ultimately yielding H_2O_2, as depicted in Eq. (4.18). Numerous studies have indicated that Mn(II)-catalyzed ozonation process is more effective in the degradation of organic compounds compared to O_3 alone (El-Raady and Nakajima, 2005; Rivas et al., 2001). The reaction pathway of Mn(II)-catalyzed ozonation process is elucidated as follows (Khuntia et al., 2016):

$$Mn^{2+} + O_3 + 2H^{+} \rightarrow Mn^{4+} + O_2 + H_2O \quad (4.21)$$

$$Mn^{4+} + 1.5O_3 + 3H^{+} \rightarrow Mn^{7+} + 1.5O_2 + 1.5H_2O \quad (4.22)$$

$$Dye + Mn^{7+} + 1.5H_2O \rightarrow Mn^{4+} + 3H^{+} + intermediates \quad (4.23)$$

$$Mn^{2+} + Mn^{4+} \rightarrow 2Mn^{3+} \quad (4.24)$$

$$Mn^{2+} + O_3 + H^{+} \rightarrow Mn^{3+} + O_2 + OH^{\bullet} \quad (4.25)$$

$$Mn^{3+} + O_3 + (Dye)^{2-} + H^{+} \rightarrow Mn^{2+} + O_2 + OH^{\bullet} + Intermediates \quad (4.26)$$

$$Dye + OH^{\bullet} \rightarrow Intermediates \quad (4.27)$$

Legube and Leitner (1999) proposed an alternative reaction pathway for Mn(II)-catalyzed ozonation process. According to their findings, an intermediate Mn^{+}-pollutant is formed, susceptible to ozone attack, leading to the oxidation of the pollutant. Similarly, Cu(II) is acknowledged as an effective homogeneous catalyst to combine with ozonation for organic pollutants degradation. Although the precise reaction pathway of Cu(II)-catalyzed ozonation process remains undisclosed, Taube and Bray (1940) have proposed a generalized reaction of metal ion-catalytic homogeneous ozonation system as follows:

$$Mn^{+n} + O_3 + H^{+} \rightarrow Mn^{+n+1} + O_2 + OH^{\bullet} \quad (4.28)$$

4.4.2 Heterogeneous Catalytic Process

Within heterogeneous catalytic processes in MBs, utilizing ozone as the gaseous medium stands out as the preferred choice. Ozone-based methods offer the advantages of a straightforward operational procedure, and given ozone's high reactivity toward numerous organic compounds and gaseous compounds in both air and water pollution, the formation of sludge is minimized. Integrating the MB ozonation process with a heterogeneous catalyst not only generates additional hydroxyl radicals but also significantly amplifies the degradation of pollutants. Another objective in employing a catalytic ozonation process is to curtail ozone consumption and subsequently reduce overall operational costs. Nonetheless, in the catalytic O_3 + MB system, the stability of the catalyst toward ozone emerges as a pivotal factor influencing process performance. Similar to organic compounds, the catalyst exhibits a propensity to degrade in an ozone environment. A schematic of a batch and continuous heterogeneous catalytic process involving MB column is shown in Figure 4.3.

Exploiting some fascinating characteristics of activated carbon such as large surface area, adsorption capacity, and chemical resistance (Faria et al., 2007), activated carbons derived from various sources such as sawdust, sewage sludge, coconut shell, and bamboo (Daud and Ali, 2004; Fang et al., 2018; Liu et al., 2010; Zhuang et al., 2014), have been studied for wastewater treatment. Enhancements in catalytic stability and activity have been achieved by incorporating metals such as Fe, Co, Mn, Al, Ni, and their oxides (Singh et al., 2022). The pivotal requirements for catalysts in catalytic ozonation, widely applicable in wastewater treatment, include high catalytic activity, facile recyclability, and reasonable cost. It is noteworthy that in the heterogeneous catalytic MB + O_3 process, the MBs facilitate increased ozone solubility in water, contributing to enhanced catalytic activity.

In their investigation, Yan et al. (2019a) explored the mechanism of Fe_2O_3, Fe_3O_4, and FeOOH in ozonation. The findings revealed that Fe_2O_3 electrostatically adsorbs ozone, consequently heightening the efficacy of O_3 decomposition into $OH^•$ radicals. Conversely, Fe_3O_4 and FeOOH demonstrated a distinct mechanism, where cyclic ionic reactions of Fe^{2+}/Fe^{3+} activated the Lewis acid sites present on the surface pores, promoting the O_3 decomposition into $O_2^{-•}$ and $OH^•$ radicals. Jothinathan et al. (2021) employed graphene oxide as a carrier for Fe_3O_4 in the catalytic O_3 process, improved O_3 mass transfer, and $^•OH$ radical generation. However, the expensive graphene oxide limited its widespread application, like water treatment. In another approach by Yan et al. (2019b), silicon particles were used as the carrier of Fe_2O_3. Fe_2O_3 and have been identified to have catalytic sites on their surface for the decomposition of O_3 into $OH^•$ radicals. But the available adsorption sites are limited for these silicon particles, which impacts the catalytic performance (Yan et al., 2019b).

Table 4.2 gathers the findings from recent investigations on heterogeneous catalytic O_3 + MB, underscoring the extensive utility of the process

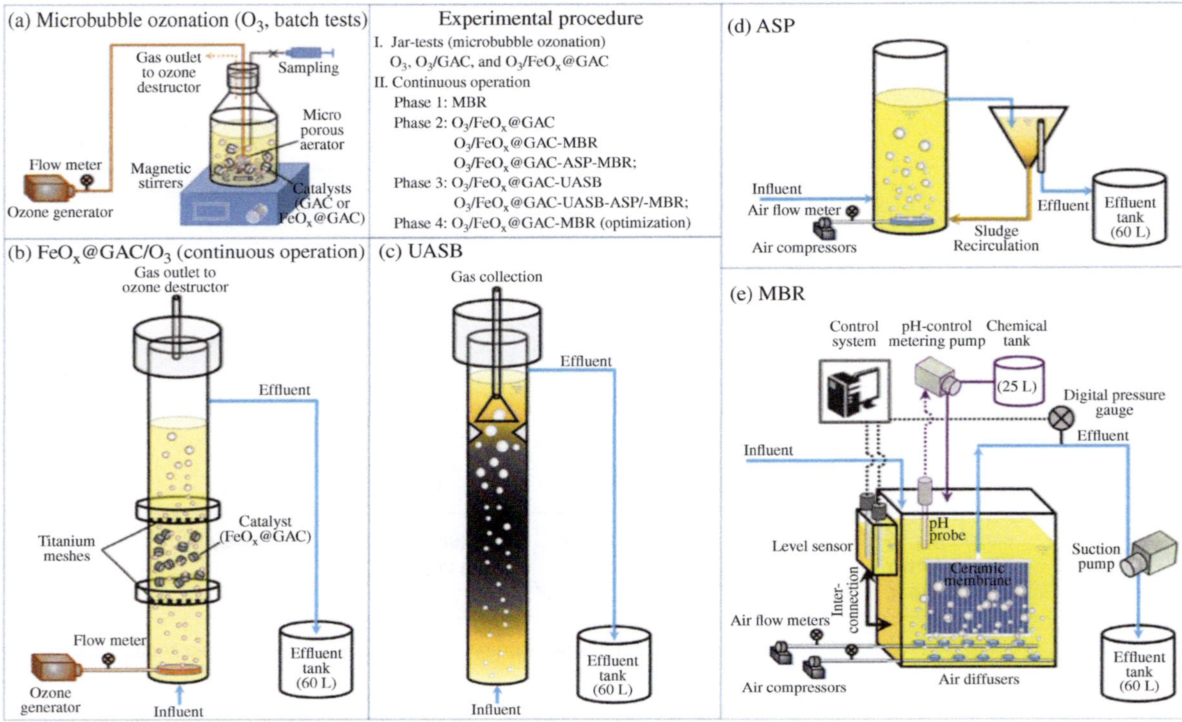

Figure 4.3 The experimental setup developed by Deng et al. for the catalytic ozonation process using MBs. Source: Deng et al. (2021)/with permission of ELSEVIER.

Table 4.2 Heterogeneous catalytic O_3 + MB process for water treatment.

Catalyst type	Wastewater	Run time (min)	O_3 dosage	Catalyst dosage (g/L)	% Enhanced by Catalytic O_3 MB system	Reference
Iron-oxides (FeOx) doped granular activated carbon	Phenolic wastewater	60	50 (mg/L)	5	30	Deng et al. (2021)
Fe and granular activated carbon	Petrochemical wastewater	100	50 (mg/L)	4	43	Jothinathan et al. (2021)
Fe_3O_4-loaded activated carbon	Pretreated coal gasification wastewater	40	0.4 L/min	3	20	Hou et al. (2020)
Activated carbon	Dye waste-water	120	0.3 L/min, 12.5 mg/min	2	20–40	Zhang et al. (2018)
Doped $MnO_2Al_2O_3$	Refractory	90	135 mg/L	70	20	Liu et al. (2021)

Source: Deng et al. (2021)/with permission of ELSEVIER.

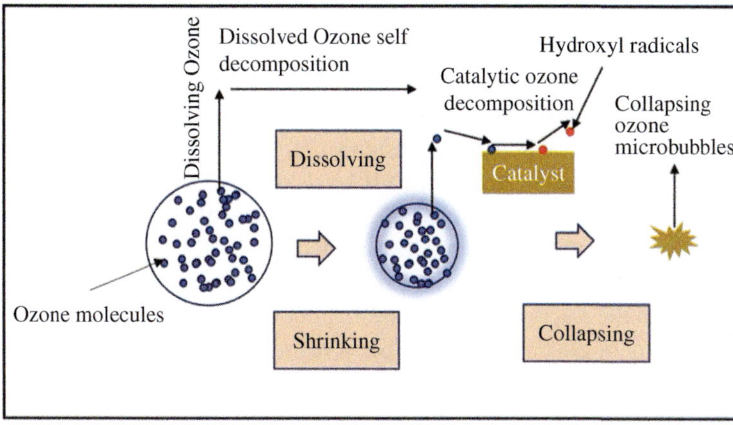

Figure 4.4 The proposed pathway for the generation of $OH^{\bullet}$ radicals in the heterogeneous catalytic O_3-MBs process. Source: Deng et al. (2021)/with permission of ELSEVIER.

in treating diverse wastewater types. Significantly, numerous studies in this domain emphasize the preference for low-cost and easily synthesized catalysts, strategically chosen to optimize operational expenses.

In a study reported by Zhang et al. (2018), a possible pathway for degradation of an azo dye pollutant in water has been proposed in a heterogeneous catalyzed MB process, shown in Figure 4.4. At first, the azo group was attacked by the $OH^{\bullet}$ radicals to form intermediate products. The amino and sulfo groups found in the azo dyes were then hydrolyzed to form either hydroxyl groups or several other intermediates. The decomposition process goes through several steps of mineralization of intermediates. Finally, the benzene ring of benzene-type acids was attacked by both O_3 and $OH^{\bullet}$ radicals to generate small organic acids. During this, the collapse of MBs accelerates the process of $OH^{\bullet}$ radical generation.

4.5 Influence of Parameters on Catalytic Processes

The parameters that significantly influence the catalytic MB process are primarily contingent on the type of catalyst employed. Among these, the pH of water emerges as a pivotal factor, governing the self-decomposition of ozone and thus standing out as the most important parameter for study. Ozone exhibits stability under acidic pH conditions and undergoes rapid dissociation. In a study by Khuntia et al. (2015), the disappearance rate of para-Chlorobenzoic acid (PCBA) was notably high in acidic conditions within the O_3 + MB system, diminishing with increasing pH. However, at pH 10, PCBA declined rapidly. It is essential to note that in the presence of other oxidants, the higher rate of oxidation may not adhere to the same pH range. The effectiveness of co-oxidants may vary, as illustrated in the accompanying Table 4.2. For instance, H_2O_2-based processes exhibit efficient pollutant removal at acidic pH levels, but it is crucial to consider that H_2O_2 also acts as a radical scavenger. Therefore, pH control is critical, and its effects must be thoroughly investigated.

Other parameters, such as gas flow rate and solid catalyst dose, are largely dependent on the reactor scale and type. In pilot-scale operations or beyond, careful control of gas flow rates is imperative for the uniform generation of MBs in the system. Although temperature is generally not a significant focus in water treatment processes regulated at 20–30 °C, temperatures exceeding 30 °C can pose challenges as gaseous O_3 becomes challenging to reside in water. Additionally, MB dissolution raises local temperatures due to the shrinking process, hindering the MB system's efficiency as higher temperatures decrease the density of bubble formation.

The concentration of H_2O_2 in conjunction with O_3 is a crucial parameter. The recommended expression for H_2O_2 dose is the molar ratio of H_2O_2 to O_3, typically advised to be in the range of 0.2–0.5. It is worth noting that the formation of superoxide ion O_2^- occurs in Eqs. (4.10) and (4.11), and as superoxide promotes ozone decomposition, the inhibition of TCE decomposition at high H_2O_2 levels raises questions.

4.6 Quantification of Radicals

4.6.1 Electron Spin Resonance Method

Prior to the quantification of free radicals, the crucial step is the detection of these radicals. The identification of free radicals, primarily hydroxyl radicals, is carried out through two separate tests, verifying the $OH^•$ radical generation resulting from the collapse of gas MBs. If O_3 is in the gas phase in the MB system, the first procedure involves ESR measurements, while the second procedure centers on the degradation of PVA as the target material. PVA, known for its resistance to O_3 degradation and limited biodegradability, serves as a suitable indicator.

Procedure 1: Many radical species exhibit moderate reactivity, often making the detection challenging within ESR time scale. To overcome this, 5,5-dimethyl-1-pyrroline-N-oxide (DMPO) is employed as a well-known spin-trap reagent. DMPO is a diamagnetic compound that can be detected by

ESR and, when reacted with the short-lived radicals, makes them stable for longer durations. Reactive oxygen species (ROS), which are otherwise undetected by ESR, are trapped in the presence of DMPO to create paramagnetic nitroxyl radical adducts, subsequently observable in ESR spectra.

In a study reported by Li et al. (2009) and Takahashi et al. (2007a, 2007b), a procedure for detecting the OH• radical was described as follows: A 5 mM DMPO solution was introduced into the water in a glass container for subsequent utilization in ESR measurements. Before the addition of DMPO, the electrical conductivity of water was measured at approximately ~1.0 μS/cm at the temperature maintained at ~20 °C. Subsequently, a water sample was transferred to a quartz flat cell after the commencement of operation by the MBs generator, and an ESR spectrum was recorded using an ESR spectrometer. The following conditions are maintained for the ESR spectrometer: room temperature, microwave power of 10 mW, modulation amplitude of 0.2 mT, 1 s time constant, and a scanning time of 16 min. Mn^{2+} is used as an external standard for the hyperfine splitting constant calibration.

The macrobubbles were generated using a glass-bonded diffuser and compared with the MBs. Figure 4.5 displays the ESR spectrum of a sample taken 20 s after the initiation of bubbling without HCl. Four lines with hyperfine splitting constants AN and AH at 14.9 G were observed, indicating that the bubbling of ozone prompted the formation of OH• radicals. The rate of OH• radical generation remained unaffected by the presence of HCl.

Figures 4.6 (a) and (b) present the outcomes of MB tests. Figure 4.6 (a) displays the ESR spectrum of a water sample captured 20 s after initiating the MBs, without HCl, while Figure 4.6 (b) depicts the ESR spectrum of a water sample containing HCl.

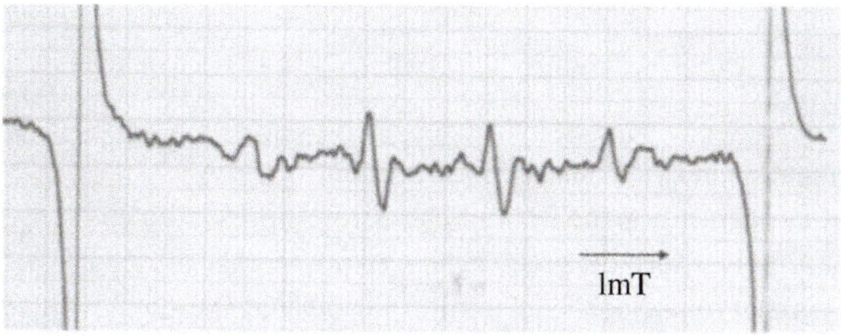

Figure 4.5 The ESR spectrum to measure the DMPO-OH in the O_3 process with macrobubbles, taken after 20 s of the beginning of the bubbles in distilled water. The spectrum is the indicative of presence of DMPO-OH, a similar spectrum observed after the addition of HCl. Source: Reproduced with permission from Takahashi 2007/American Chemical Society.

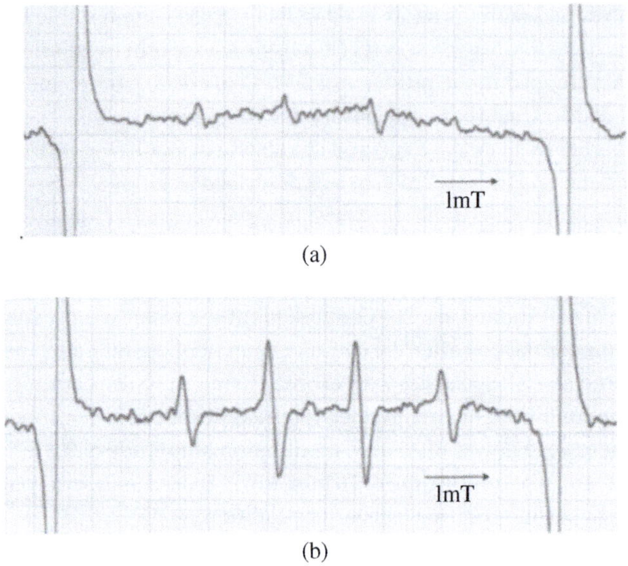

Figure 4.6 The ESR spectrum to measure the DMPO-OH in the O_3 process with MBs: (a) taken after 20 s of the beginning of the bubbles in distilled water, (b) after 5 s using HCl. The spectrum is the indicative of presence of DMPO-OH. Reproduced with permission from Takahashi 2007/American Chemical Society.

Procedure 2: The decomposition of PVA was examined with and without MBs. PVA does not show any reaction with O_3 only but reacts with $OH^•$ radicals. However, when used with O_3-MBs, showed a declined concentration with time. This was also confirmed with the detection of TOC, owing to a conclusion that the generation of $OH^•$ radicals takes place in O_3-MB system. This method is known as an indirect method for the detection of $OH^•$ radicals in water.

4.6.2 R_{ct} Method

4.6.2.1 Determination of $OH^•$ Radical Concentration

Determining the concentration of $OH^•$ radicals directly presents a challenge due to their extremely short lifetime (a few nanoseconds of order) and their tendency to react onsite of the formation (Roots and Okada, 2014). To overcome this difficulty, an indirect methodology was proposed by Elovitz and von Gunten (1999) for measuring $OH^•$ radical concentrations. In this approach, a probe compound, such as PCBA, is employed that is highly reactive with $OH^•$ radicals but shows no reactivity toward O_3 (Pi et al., 2005; Rivas et al., 2005). Notably, the rate constants of PCBA with $OH^•$

radicals and O_3 are 5×10^9 and 0.15 L/mol·s, respectively. When both PCBA and a scavenger are there in water, the consumption of $OH^\bullet$ radicals occurs due to both entities.

At extremely low concentrations of PCBA, the impact of the probe on scavenging $OH^\bullet$ radicals is minimal, resulting in a negligible contribution to the overall scavenging process. In such cases, the rate equation for PCBA, which exhibits significantly less reactivity with O_3 compared to $OH^\bullet$ radicals, can be expressed as per the formulation by Elovitz and von Gunten (1999).

$$-\frac{d[PCBA]}{dt} = k_h[OH^\bullet][PCBA] \tag{4.29}$$

$$-ln\left(\frac{[PCBA]}{[PCBA]_0}\right) = k_h \int [OH^\bullet]dt \tag{4.30}$$

where $\int [OH^\bullet]dt$, signifies the time integrated of $OH^\bullet$ radical concentration, denoted as $OH^\bullet$ exposure.

Equation (4.30) reveals that the reduction in the PCBA concentration with time, t, serves as an indirect measure of the $OH^\bullet$ radical exposure during the reaction. Consequently, a factor denoted as R_{ct} is introduced, as follows:

$$R_{ct} = \frac{\int [OH^\bullet]dt}{\int [O_3]dt} \tag{4.31}$$

The O_3-exposure, or $\int [O_3]dt$, is determined by computing the integration of the O_3 concentration profile, as illustrated in Figure 4.7. Utilizing Eqs. (4.30) and (4.31),

$$\ln\left(\frac{[PCBA]}{[PCBA]_0}\right) = -k_h R_{ct} \int [O_3]dt \tag{4.32}$$

Equation (4.32) reveals that R_{ct} can be ascertained by monitoring the variations in the concentrations of both PCBA and $OH^\bullet$ radicals. The $OH^\bullet$ radicals from O_3 are utilized in the reaction with PCBA, causing a decrease in PCBA concentration. Simultaneously, O_3 dissolution in water increases, leading to a rise in $\int [O_3]dt$. A linear plot of Eq. (4.32) between $n\left(\frac{[PCBA]}{[PCBA]_0}\right)$ and $-k_h \int [O_3]dt$, the slope R_{ct} can be derived, as depicted in Fig. 4.8.

The O_3 exposure was notably higher under acidic pH conditions than alkaline pH. This difference is attributed to the enhanced O_3 self-decomposition at alkaline pH. Consequently, molecular ozone exhibits greater stability at acidic pH levels. The correlation of $\ln\left(\frac{[PCBA]}{[PCBA]_0}\right)$ and O_3 exposure shows that R_{ct} is a time-independent constant. Elovitz and von Gunten (1999) have expressed R_{ct} as:

$$R_{ct} = \frac{[OH^\bullet]}{[O_3]} \tag{4.33}$$

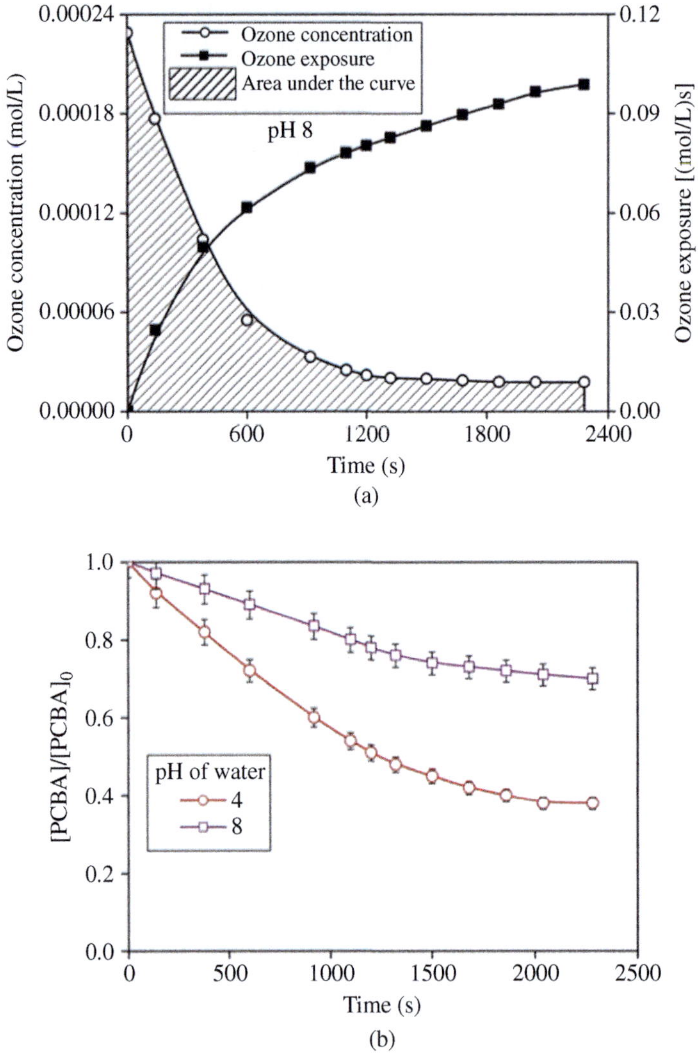

Figure 4.7 (a) The O_3 concertation and exposure profile at pH 8, and (b) variation of the PCBA probe compound concentration. Source: Khuntia et al. (2015)/with permission of ELSEVIER.

The apparent $[OH^{\bullet}]$ is measured from Eq. (4.33). The pH-dependent profile of $OH^{\bullet}$, O_3 exposure, and R_{ct} is shown in Fig. 4.9.

Elovitz and von Gunten (1999) observed an increase in R_{ct} with rising pH, while $OH^{\bullet}$ exposure values remained closely aligned within the pH

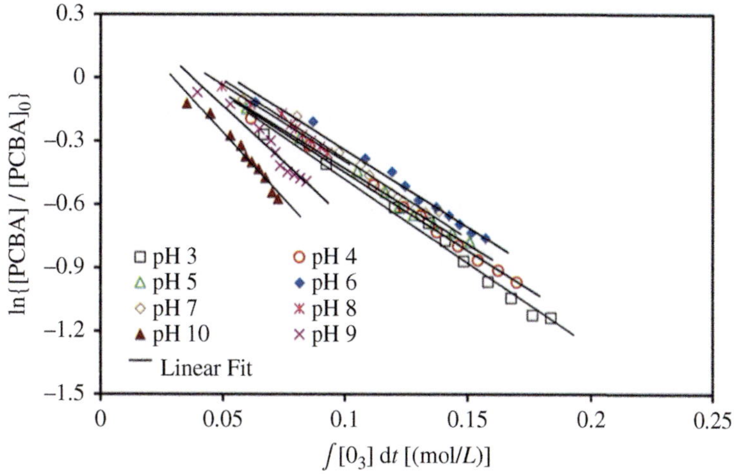

Figure 4.8 The leaner fit to obtain the value of R_{ct} using Eq. (4.32). Source: Khuntia et al. (2015)/with permission of ELSEVIER.

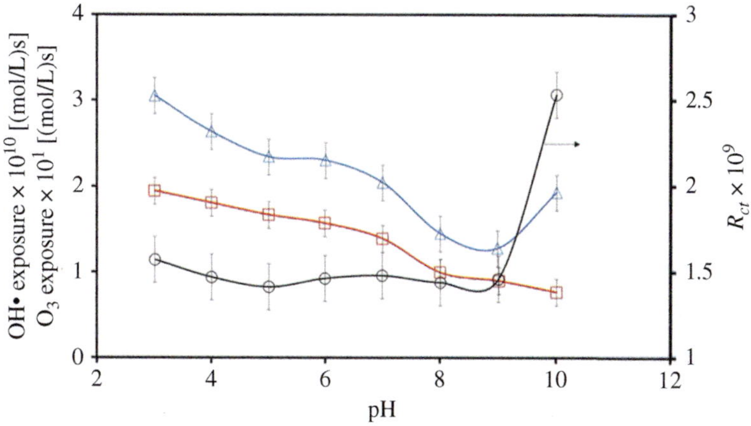

Figure 4.9 The profile of O_3, $OH^•$ radicals exposure, and R_{ct} as a function of pH. Source: Khuntia et al. (2015)/with permission of ELSEVIER.

range of 6 to 9 (specifically, at 5×10^{-14} mol/L s). This convergence was attributed to heightened O_3 decomposition and reduced $OH^•$ radical generation within this pH range. It was reported by Haag and Yao (1993) that $[OH^•]$ remained nearly constant at pH 7–8.2 across various sources of water samples.

4.6.2.2 Evaluate the Contribution of OH• Radical for Pollutant Degradation

It is believed that OH• radicals and molecular O_3 react simultaneously with the water micropollutants. Therefore, the rate of disappearance of the pollutant (Y) with O_3 can be written as:

$$-\frac{d[Y]}{dt} = k_h[OH^\bullet][Y] + k_{O_3}[O_3][Y] \tag{4.34}$$

By introducing Eq. (4.33) in Eq. (4.44)

$$-\frac{d[Y]}{dt} = k_h R_{ct}[O_3][Y] + k_{O_3}[O_3][Y] \tag{4.35}$$

$$-\frac{d[Y]}{dt} = (k_h R_{ct} + k_{O_3})[O_3][Y] \tag{4.36}$$

where k_h and k_{O_3} are the 2nd-order rate constants for the reaction of the micropollutant with OH• radicals and O_3.

The R_{ct} parameter is also used for evaluation of the contributions of OH• radicals and O_3 in the degradation process. The fraction of Y degraded by OH• and O_3 is expressed as

$$f_{OH^\bullet} = \frac{\text{Part of } Y \text{ reacted with OH}^\bullet}{\text{Part of } Y \text{ reacted with OH}^\bullet \text{ radicals} + \text{part of } Y \text{ reacted with O}_3} \tag{4.37}$$

$$f_{OH^\bullet} = \frac{k_h[OH^\bullet][Y]}{k_h[OH^\bullet][Y] + k_{O_3}[O_3][Y]} \tag{4.38}$$

rearranging the above equations,

$$f_{OH^\bullet} = \frac{k_h R_{ct}}{k_h R_{ct} + k_{O_3}} \tag{4.39}$$

Using Eq. (4.49), the fraction of a target compound reacted with OH• radicals can be calculated.

Nomenclature

Notation		SI Unit
d_b	diameter of the bubble	m
k	rate constant	1/mol.s
k_h	rate constant for oxidation of the probe compound	1/mol.s
$[O_3]$	concentration of O_3	mol/L

$[OH^\bullet]$	concentration of $OH^\bullet$ radicals	mol/L
pK	dissociation constant	unit less
P	pressure inside the bubble	pa
P_G	gas pressure	pa
P_L	liquid pressure	pa
[PCBA]	concentration of PCBA	mol/L
$[PCBA]_0$	initial concentration of PCBA	mol/L
R_{ct}	ratio of $OH^\bullet$ and O_3 exposure	
t	time	s
Y	target pollutant in water	
$[Y]$	concentration of pollutant Y	mol/L
	Greek Nomenclature	
σ	surface tension	N/m

References

Azuma, T., Otomo, K., Kunitou, M. et al. (2019). Removal of pharmaceuticals in water by introduction of ozonated microbubbles. Sep. Purif. Technol. 212: 483–489.

Dalal P., Dalai S., Khuntia S., Experimental study of the removal NO_x and SO_2 from flue gas using the O_3/H_2O_2, and O_3/UV processes, Environ. Tech. 2023. doi: https://doi.org/10.1080/09593330.2023.2256991.

Daud, W.M.A.W. and Ali, W.S.W. (2004). Comparison on pore development of activated carbon produced from palm shell and coconut shell. Bioresour. Technol. 93: 63–69.

Deng, S., Jothinathan, L., Cai, Q. et al. (2021). FeOx@GAC catalyzed microbubble ozonation coupled with biological process for industrial phenolic wastewater treatment: catalytic performance, biological process screening and microbial characteristics. Water Res. 190: 116687.

El-Raady, A.A. and Nakajima, T. (2005). Composition of carboxylic acids in water by O_3, O_3/H_2O_2, and O_3/catalyst. Ozone Sci. Eng. 27: 11–18.

Elovitz, M.S. and von Gunten, U. (1999). Hydroxyl radical/ozone ratios during ozonation processes. I. the R_{ct} concept. Ozone Sci. Eng 21: 239–260.

Fang, X.Y., Yang, M., Li, J.X. et al. (2018). Preparation of bamboo activated carbon by microwave radiation with water vapour activation and investigation of its adsorption properties. Desalin. Water Treat. 135: 351–361.

Faria, P.C.C., Orfao, J.J.M., and Pereira, M.F.R. (2007). Ozonation of aniline promoted by activated carbon. Chemosphere 67: 809–815.

Glaze, H.W., Kang, J.W., and Chapin, H.D. (1987). The chemistry of water treatment processes involving ozone, hydrogen peroxide and ultraviolet radiation. Ozone. Sci. Eng. 9: 335–352.

Haag W.R., Yao C.C.D. Ozonation of US drinking water sources: OH concentration and oxidation-competition values. In: ozone in water and wastewater treatment, Proc. 11th Ozone World Congress. 1993; 2: 119–126.

Hou, S., Jia, S., Jia, J. et al. (2020). Fe_3O_4 nanoparticles loading on cow dung-based activated carbon as an efficient catalyst for catalytic microbubble ozonation of biologically pretreated coal gasification wastewater. J. Environ. Manag. 267: 110615.

Jothinathan, L., Cai, Q.Q., Ong, S.L., and Hu, J. (2021). Organics removal in high strength petrochemical wastewater with combined microbubble-catalytic ozonation process. Chemosphere 263: 127980.

Khuntia, S., Majumder, S.K., and Ghosh, P. (2016). Catalytic ozonation of dye in a microbubble system: hydroxyl radical contribution and effect of salt. J. Environ. Chem. Eng. 4: 2250–2258.

Khuntia, S., Majumder, S.K., and Ghosh, P. (2015). Quantitative prediction of generation of hydroxyl radicals from ozone microbubbles. Chem. Eng. Res. Des. 98: 231–239.

Khuntia, S. and Mohan, G. (2022). A comparative study of the catalytic and non-catalytic ozone based processes for simultaneous of SO_2 and NO_X removal. J. Ind. Eng. Chem. 106: 152–159.

Khuntia, S., Sinha, M.K., and Mohan, G. (2020a). Evaluation of reaction kinetics for removal of NO_x by ozone and hydrogen peroxide. Ind. Eng. Chem. Res. 59: 17806–17814.

Khuntia, S., Sinha, M.K., and Saini, B. (2020b). Conversion of NO_2 through ozonation and peroxone process in gas and aqueous phase: Finding the suitable process through experimental route. Chem. Eng. J. 387: 124082.

Khuntia, S., Sinha, M.K., and Singh, P. (2021). Theoretical and experimental investigation of the mechanism of the catalytic ozonation process by using a manganese-based catalyst. Environ. Technol. 42: 632–639.

Legube, B. and Leitner, N.K. (1999). Catalytic ozonation: a promising advanced oxidation technology for water treatment. Catal. Today. 53: 61–72.

Li, P., Takahashi, M., and Chiba, K. (2009). Degradation of phenol by the collapse of microbubbles. Chemosphere 75: 1371–1375.

Liu, X., Wang, S., Yang, H. et al. (2021). Characterization of a doped $MnO_2Al_2O_3$ catalyst and its application in microbubble ozonation for quinoline degradation. Ozone. Sci. Eng. 43: 173–184.

Liu, Z.G., Zhang, F.S., and Sasai, R. (2010). Arsenate removal from water using Fe_3O_4-loaded activated carbon prepared from waste biomass. Chem. Eng. J. 160: 57–62.

Moriwaki, H., Takagi, Y., Tanaka, M. et al. (2005). Sonochemical decomposition of perfluorooctane sulfonate and perfluorooctanoic acid. Environ. Sci. Technol. 39: 3388–3392.

Pi, Y.M., Ernst, J.C., and Schrotter. (2003). Effect of phosphate buffer upon CuO/Al_2O_3 and Cu (II) catalyzed ozonation of oxalic acid solution. Ozone Sci. Eng. 25: 393–397.

Pi, Y., Schumacher, J., and Jekel, M. (2005). The use of para-chlorobenzoic acid (pCBA) as an ozone/hydroxyl radical probe compound. Ozone Sci. Eng. 27: 431–436.

Prengle, H.W.J.R., Mauk, C.E., Legan, R.W., and Hewes, C.G. (1975). Ozone/UV process effective wastewater treatment. Hydrocarbon Processing 54: 82–87.

Rivas, F.J., Beltrán, F.J., Acedo, B. et al. (2005). Kinetics of the ozone-p-chlorobenzoic acid reaction. Ozone Sci. Eng. 27: 3–9.

Rivas, J., Rodriguez, E., Beltran, F.J. et al. (2001). Homogeneous catalyzed ozonation of simazine: effect of Mn (II) and Fe (II). J. Environ. Sci. Health B. 36: 317–330.

Rodriguez-lopez, A.D., Garcia-garrido, J., Perez-ramiro, C., and Garcia-castello, E.M. (2013). discoloration on methylene blue solutions by direct and catalytic ozonation. J. Mater. Sci. Chem. Eng. 1: 33–38.

Roots, R. and Okada, S. (2014). Estimation of lifetimes and diffusion distances of radicals involved in X-ray-induced DNA strand breaks or killing of mammalian cells. Radiat. Res. 64: 306–320.

Singh, A., Sinha, M.K., and Khuntia, S. (2022). Enhanced generation of hydroxyl radicals by OMS-2 catalyst for flue gas absorption. Mater. Today Proc. 57: 279–284.

Takahashi, M. (2005). ζ potential of microbubbles in aqueous solutions: electrical properties of the gas-water interface. J. Phys. Chem. B 109: 21858–21864.

Takahashi, M., Chiba, K., and Li, P. (2007a). Formation of hydroxyl radicals by collapsing ozone microbubbles under strongly acidic conditions. J. Phys. Chem. B 111: 11443–11446.

Takahashi, M., Chiba, K., and Li, P. (2007b). Free-radical generation from collapsing microbubbles in the absence of a dynamic stimulus. J. Phys. Chem. B 111: 1343–1347.

Takahashi, M., Ishikawa, H., Asano, T., and Horibe, H. (2012). Effect of microbubbles on ozonized water for photoresist removal. J. Phys. Chem. C 116: 12578–12583.

Takahashi, M., Kawamura, T., Yamamoto, Y. et al. (2003). Effect of shrinking microbubble on gas hydrate formation. J. Phys. Chem. B. 107: 2071–2173.

Bray, T.W.C. (1940). Chain reactions in aqueous solutions containing ozone, hydrogen peroxide and acid. J. Am. Chem. Soc. 97: 189–192.

Tsuge, H. (2014). *In Micro- and Nanobubbles: Fundamentals and Applications.* Boca Raton, FL: CRC Press.

Wu, C.H., Kuo, C.Y., and Chang, C.L. (2008). Homogeneous catalytic ozonation of C.I. Reactive Red 2 by metallic ions in a bubble column reactor. J. Hazard. Mater. 154: 748–755.

Yan, L., Bing, J., and Wu, H. (2019a). The behavior of ozone on different iron oxide surface sites in water. Sci. Rep. 9: 14752–14710.

Yan, P., Shen, J., Yuan, L. et al. (2019b). Catalytic ozonation by Si-doped α-Fe_2O_3 for the removal of nitrobenzene in aqueous solution. Sep. Purif. Technol. 228: 115766.

Yasui, K., Tuziuti, T., and Kanematsu, W. (2018). Mysteries of bulk nanobubbles (ultrafine bubbles); stability and radical formation. Ultrason. Sonochem. 48: 259–266.

Zhang, J., Huang, G.-Q., Liu, C. et al. (2018). Synergistic effect of microbubbles and activated carbon on the ozonation treatment of synthetic dyeing wastewater. Sep. Purif. Technol. 201: 10–18.

Zhuang, H.F., Han, H.J., Hou, B.L. et al. (2014). Heterogeneous catalytic ozonation of biologically pretreated Lurgi coal gasification wastewater using sewage sludge based activated carbon supported manganese and ferric oxides as catalysts. Bioresour. Technol. 166: 178–186.

5

Water Treatment and Related Applications of Microbubbles

Dr Manish Kumar Sinha

Associate Professor, Department of Chemical Engineering, Pandit Deendayal Energy University, Gandhinagar, Gujarat, India

5.1 Introduction

Looking at the application of microbubbles (MBs), there are three major areas where the fantastic properties of the MBs have been utilized, such as enhanced drug delivery for medical/therapeutic applications, water treatment, and mineral beneficiations. Many research articles have been published on the application of MBs and integrative methods for water treatment. Not only that, numerous review articles have been published worldwide that have meticulously gathered information on the application of MBs across laboratory, pilot, and industrial scales. It is impossible to cite all the work here, but to name some, the readers can follow Agarwal et al. (2011); Khuntia et al. (2012); Kaushik and Chel (2014); Tekile et al. (2017); John et al. (2020); Selihin and Tay (2021); Nair et al. (2022); Zhou et al. (2022); and Levitsky and Tavor, (2022).

On an industrial scale, the water treatment process has a lot of scope for improvement in order to achieve sustainability goals. Modern water treatment technologies are working on cleaner processes that are less energy-intensive and inexpensive. In search of these criteria, the processes are also expected to produce less sludge and be versatile in nature to handle multiple types of pollutants under a wide range of operating parameters. In most parts of the world, the traditional process of water treatment consisting of the steps shown in Figure 5.1 is still followed. It is rather challenging to completely replace the existing process, which could be an expensive route to follow. Therefore, another challenge in the advanced or any new technology is the partial replacement or integration of the technology

Microbubbles: Engineering Aspects and Industrial Applications, First Edition. Snigdha Khuntia.

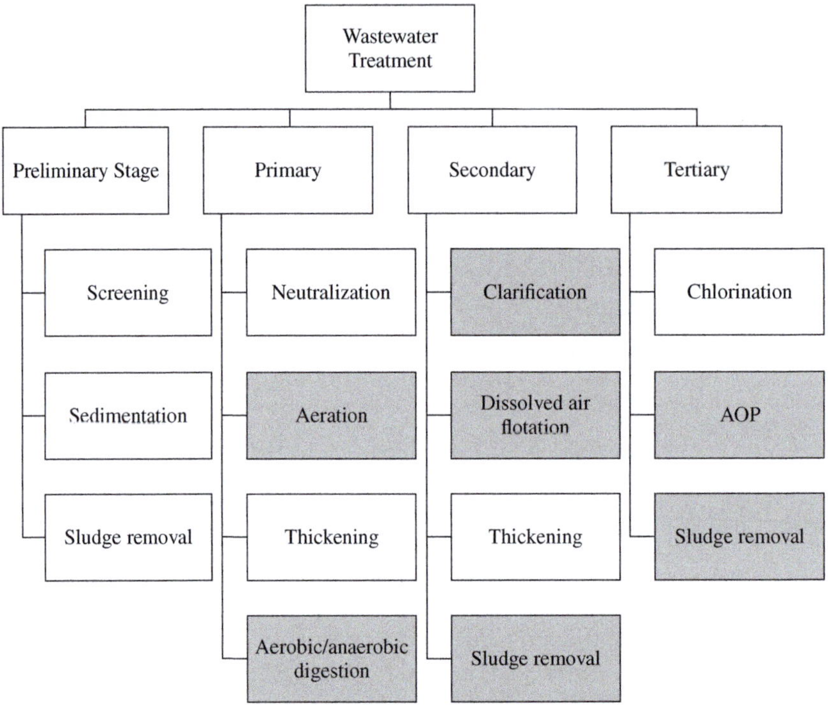

Figure 5.1 The stages of water treatment and the scope of integration of MB-based techniques studied so far from the literature. The stages in gray are those reported in literature where the MB systems have been integrated to treat the industrial wastewater at different scales of operation.

with the existing method followed by the plants. Most of the processes are either pH-sensitive (e.g. coagulation), pollutant-type selective in nature (oxidative methods), make a lot of sludge (adsorption-based processes, Fenton processes), or have chances of failure due to clogging (e.g. membrane processes), or are energy-intensive (plasma-based oxidative methods). The water treatment plants are always looking for techniques that are easy to operate and do not have a lot of limitations in terms of process parameters. Looking at the severity of global environmental challenges, each year the environmental norms for discharging wastewater to the water bodies and disposal parameters are getting stricter for each country. The industries, at the same time, are under a lot of pressure to modernize their existing technology but are struggling to find a suitable technology that could meet the government norms as well as their operation and capital costs.

In this context, the MBs have shown a great potential in water treatment at several stages of water treatment where the fantastic properties of the MBs can be utilized to improve the effectiveness of the overall wastewater treatment process. The stages (highlighted as gray) are the operational stages where numerous researchers have tested the effectiveness of MBs coupled with newer or existing methods. In this chapter, the potential of MBs in water treatment at various stages has been discussed.

5.2 Preliminary and Primary Stages of Water Treatment

The conventional wastewater treatment plant (WWTP) is designed to treat the bigger suspended solids and heavy total dissolved solids (TDS) loads of industrial wastewater at the primary stage. The preliminary stage mostly works by the gravitational force to settle down the TDS and suspended particles, or sometimes uses screening methods for settling the solids. The sludge formation at the end of this stage is high. Only at the sedimentation stage, there is scope for improvement to make the process faster. For industrial raw wastewater, there is no scope for utilization of MBs at this stage. In the case of river and pond raw water, hospital waste, and winery wastewater, the treatment processes begin from the primary stage, where the amount of TDS is less and the particle size of the suspended solids is less compared to industrial sources. Therefore, for these sources of wastewater, the treatment process begins with neutralization, where the pH is adjusted close to 6–8, using either an acid or alkali. After this step, the process of aeration begins, where the properties of MBs have a great potential to enhance aeration efficiency and are hence studied by many researchers (follow Table 5.1).

5.2.1 Aeration Process

The traditional aeration process simply uses milli-bubbles or macrobubbles to dissolve oxygen or air to facilitate further in the biologically activated sludge tank. The primary role of the aeration tank is to dissolve more oxygen into the tank to promote the degradation of any organic material, provide a good condition for bacterial growth, and provide longer residence time for the organic pollutants to break down. The higher the mass transfer of the gases in water, the higher would be the digestion of the pollutants in further processes. Moreover, it is always taken care that the O_2 level in the aeration tank is higher than 2 mg/L. If the O_2 levels go below 2 ppm, the bacterial growth stops. However, due to lower solubility of oxygen and lower surface area of the bubbles used in this process, maintaining an optimum O_2 level poses a challenge in this step. In an aeration tank, when MBs are used, surprisingly not only the higher mass transfer of oxygen was achieved, but also the MBs were able to float the residual suspended particles through the

Table 5.1 Application of Microbubbles (MBs) in the primary stage of wastewater treatment (work in 2010–2024).

	Aeration process							
Wastewater type	**MB size (μm)**	**Gas phase**	**Flow rate/ Dosage**	**Temperature (°C)**	**pH**	**BOD; COD (mg/L or %)**	**Scale**	**Reference**
Synthetic	2.6	air	0.67 L/min	18		600	Pilot (50 L)	Yao et al. (2016)
River raw water	50–100	air	0.1 L/min			0.32–0.95; 11–21	Laboratory	Ali et al. (2021)
Ponds raw water	0.1–0.2	O_3	1.0 mg/L		7.3	63–91%; 55–82%	Pilot	Pal et al. (2022)
Hospital waste	0.02–4	O_3		21.5	7.4 ± 0.3	90%; 92%		Menendeza and Floresa (2017)
Industrial wastewater	0.4–3	O_3	1 L/min	20	3 to 11	63%	Laboratory	Xia and Hu (2019)
River water	0.25	O_3		22.7	7 ± 0.2	69.36%	Pilot	Salgueroa and Flores et al. (2017)
Raw municipal waste			2.5 Lpm	25–35	7.72		Pilot	Farajzadehha et al. (2022)
Synthetic	0.04–500	O_2/ CO_2	$10–60 \times 10^{-6}$	25–27			Laboratory	Fan et al. 2010
Groundwater			0.5–1.5 Lpm			1.92	Laboratory	Suwartha et al. (2020)
Textile wastewater		O_3, air	0.02–1.5 Lpm	30–33	7–9	80–120; 500–600	Laboratory	Chua et al. (2008)
Winery wastewater	15	O_3	3.6–5.4 Lpm		4–10		Pilot	Lucasa et al. (2009)
Aerobic/anaerobic digestion								
Fiber Industry	45	O_3	0.5 L/min	6.5 to 8.3	20	42%	Laboratory	Zheng et al. (2015)
Real bio-treated coal chemical wastewater	<50	O_3	2 Lpm	26.7 C	8	0.038; 283.8	Laboratory	Liu et al. (2018)

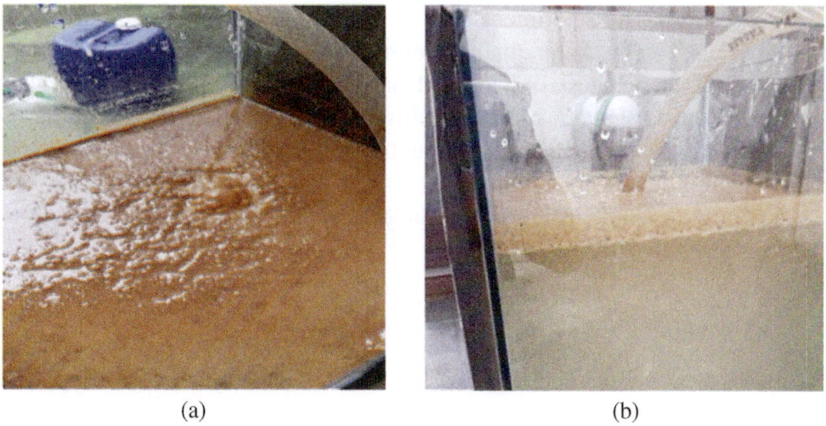

Figure 5.2 MBs used in the aeration of wastewater (a) a photograph of top view and (b) a photograph of side view. Source: Reproduced with permission from Ali et al. (2021)/IOP Publishing/CC BY 3.0.

enhanced flotation process to the surface level, which can be scaped out and separated. In a study reported by Ali et al. (2021) for contaminant removal of smaller suspended particles and to get higher oxygen solubility at the aeration tank, a pilot-scale operation was performed (as shown in Figure 5.2)

In an interesting study by Pal et al. (2022), the pond water was subjected to MBs in the city of Mumbai, India, to treat the pond water in situ with air and O_3 nanobubbles of 100–200 nms (follow Figure 5.3). Their study reported the changes in various water parameters such as biological oxygen demand (BOD), chemical oxygen demand (COD), and total soluble

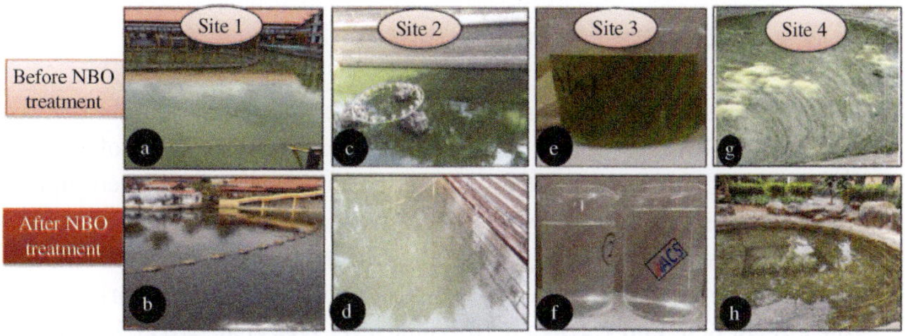

Figure 5.3 The photographs of MB- and NB-based water treatment for water quality improvement in ponds at various locations in Mumbai, India. Source: Reproduced with permission from Pal et al. (2022)/ELSEVIER.

solids (TSS). The experiments were carried out at different seasons to check the change in the water quality parameters throughout the year.

In a similar context, Li-Bang et al. have treated textile wastewater through an enhanced aeration system by MBs, where the volumetric mass transfer rate of oxygen was achieved at about 0.086–0.413 kg/m^{3}h, along with a total mass transfer coefficient of 0.1072–0.4859 1/min that could be obtained at 0.02–1.5 dm^3/min airflow rates. The MBs helped to get a 20% higher COD removal rate.

5.2.2 Aerobic/Anaerobic Digestion Process

Aerobic and anaerobic digestion, as their name says, are two types of biological treatment processes used for wastewater treatment. The two treatment processes can be used separately or together in a plant depending. Both of the processes use microorganisms to mineralize the organic content of water, and the widely used parameters to detect their potential are BOD and COD. The major difference between these two processes is their oxygen requirement, where the aerobic process does need oxygen for the growth of bacteria and further the digestion of organic pollutants, whereas the anaerobic process works without oxygen. The combined aerobic and anaerobic system is often used for the treatment of heavy organic content wastewater. The anaerobic system reduces the initial organic content to a large extent, followed by the aerobic system to further bring down the BOD level to the lowest possible value. The aerobic system is also used for ammonia and odor removal to form nitrate. A combination of aerobic and anaerobic systems is rather expensive but is capable of complete mineralization of organic pollutants in wastewater.

5.3 Secondary Effluent Treatment

The primary and secondary steps of wastewater treatments remove the vast majority of BOD and TSS in wastewater. However, the presence of residual dissolved organic matter (DOM) in the secondary effluent is often high in amount (refer to Table 5.2). Modern WWTPs even use ultrafiltration, microfiltration-based membrane processes, adsorption, chlorine, and O_3-based oxidation process after the secondary treatment step, to further bring down the level of BOD and TSS to a permissible level. (Bai et al., 2017). Membrane filtration is very commonly used for the residual TSS removal from the secondary effluents. A list of work from the literature on the application of MB in secondary treatment processes have been depicted in Table 5.3.

Table 5.2 The characteristics of real effluents before and after MB treatment.

	pH	Eh (mv)	T (°C)	NH_3-N (mg/L)	NO_3^- (mg/L)	OCN^- (mg/L)	Cu (mg/L)	Fe (mg/L)	Mn (mg/L)	Zn (mg/L)
Before treatment										
stream 1	9.5	260	10	22.3	59	20.7	20.2	0.29	0.03	3.43
stream 2	9.0	242	22	32.3	173	<0.05	<0.05	<0.05	2.22	0.40
After treatment										
stream 1	8.7	350	27	16.1	125	<0.05	0.54	0.06	0.02	5.74
stream 2	9.6	534	36	0.2	275	<0.05	<0.05	<0.05	<0.001	0.22

Source: Ryskie et al. (2020)/with permission of ELSEVIER.

In the municipal wastewater treatment studied by Fan et al. (2019), a combination of MB-UV-TiO_2 was used where the shrinking MBs were effective in the generation of reactive radicals under UV irradiation. The experimental setup developed by them is shown in Figure 5.4. The reactor consists of a UV lamp of 40 W coated with TiO_2/Ni film, placed at the core of the reactor. The reactor column has a volume of 8 L, a height of 100 cm, and an inner diameter of 12 cm. The water flows through the photocatalytic reactor upwards and circulates between the MBs tank, which is a separate one.

The mechanism of the catalytic MB-UV-TiO_2 was proposed by Fan et al., as shown in Figure 5.5. The collapse of MBs, which is inversely proportional to the diameter of MBs, creates larger internal pressure due to its smaller size. This principle follows the Young–Laplace equation while shrinking. As the bubble shrinks, the temperature inside the bubble also rises rapidly, which is also dependent on the pH of the water. The shrinkage of the bubble and rise in temperature are accelerated by external stimuli such as UV irradiation. It is reported that the MBs are negatively charged across a wide range of pH and can attribute different charges and ζ potential by changing the water pH.

5.3.1 Mining Wastewater

The treatment of mining wastewater is a challenging process that is high in suspended solids containing minerals. The mining industry has several types of contaminants including NH_3-N, high sludge content, and high TDS. Due to the high salinity of mineral wastewater, the biological and some advanced oxidation process (AOPs) do not show good efficacy. In a study by Ryskie et al. (2020), the O_3-MBs are very effective in reducing the

Table 5.3 MB application in the secondary treatment process (work in 2010–2024).

Wastewater type	MB size (μm)	Gas phase	Temperature (°C)	pH	BOD; COD (mg/L or %)	Scale	Reference
Municipal	25	Air	20	7.5–8.44	44.64%	Laboratory	Fan et al. (2019)
Dye wastewater	<45	air	23	9	COD: 89% removal	Laboratory	Liu et al. (2010)
Pharmaceutical and personal care product (PPCP)		O_3	37		7.28	Laboratory	Kim and Lee (2022)
Full-scale plant in Japan	45	air	23	7.6	1–1.5 mg/L	Full scale	Hashimoto et al. (2021)
Mineral	<45	O_3	25	8	68%; 87.6%	Laboratory	Zheng et al. (2015)

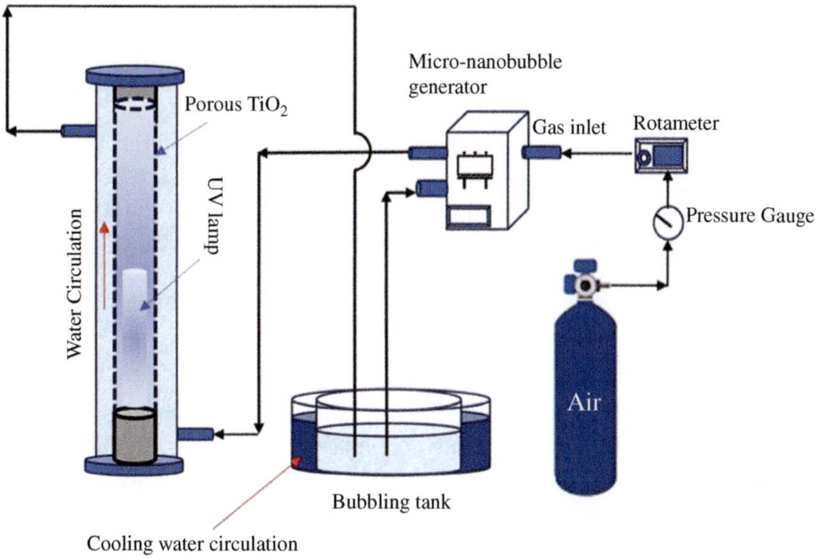

Figure 5.4 The catalytic UV-MB-based experimental setup. Source: Fan et al. (2019)/with permission of ELSEVIER.

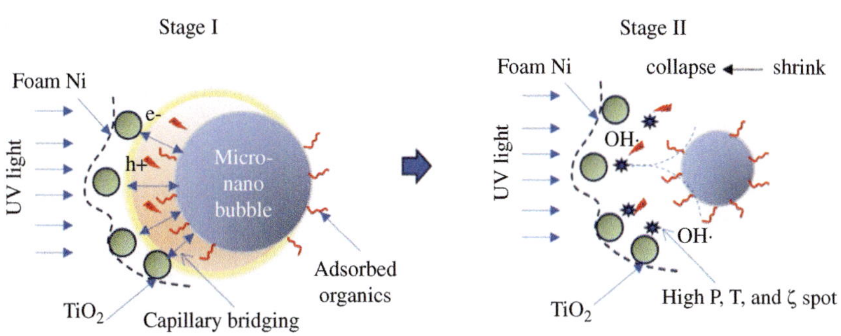

Figure 5.5 The mechanism of collapse of MBs in the presence of UV irradiation. Source: Fan et al. (2019)/with permission of ELSEVIER.

NH_3-H content to 99%, for mining wastewater. Table 5.2 lists the parameters tested for the real effluent before and after the treatment using MBs.

5.3.2 Pharmaceutical and Personal Care Products (PPCPs) Effluents

Pharmaceutical and personal care products (PPCPs) effluents are considered one of the most hazardous wastewater sources for the WWTPs for treatment. Numerous laboratory-scale studies have been carried out utilizing AOPs, nanoparticle-based processes, membrane technology, and integrated systems (Singh et al. 2022, 2024). MBs have been tried coupled with O_3, H_2O_2, or some more oxidative and AOP routes for the treatment of PPCPs effluents. Some generic compounds found in the secondary effluent of PPCP are shown in Figure 5.6 (a). Kim and Lee (2022) have used O_3-MBs for the removal of secondary effluents from PPCPs wastewater and found the method very promising for the removal of 26 PPCPs simultaneously.

5.3.3 Sludge Removal Step

Biological WWTPs produce excessive sludge and additional operations. Ozonation process is a good methodology for reducing sludge formation and enhancing sludge disintegration. MBs when coupled with O_3 have proven to be a good method for sludge reduction from the secondary effluents. Hashimoto et al. (2021) have demonstrated the schematic diagram for a full-scale WWTP using a spiral liquid flow MBs generator. The water treatment plant worked as per the conventional processes, but at the sludge removal step, the MBs have been used to improve the quality of the sludge tank. The schematic of the WWTP is shown in Figure 5.7. At 15 mg/L O_3, 92% sludge reduction was achieved by Hashimoto et al. (2021), which was studied for a full-scale WWTP, and the various water quality parameters at different stages are compared in Table 5.4.

Table 5.4 The water parameters of the influent and effluents of wastewater treatment plant using O_3-MBs treatment process.

Operation		Flow rate m^3/d	T (°C)	pH	BOD (mg/L)	SS (mg/L)	TN (mg/L)	TP (mg/L)
Pre-operation	Influent Effluent	48.8	21.9	7.6	8.0	790 ± 310	6.9 ± 310	
2nd aeration tank								
O_3 period	Influent Effluent	46.5	24.6	7.6	5.8	390 ± 160	7.7 ± 7.8	3.9 ± 1.2
No O_3 period	Influent Effluent	48.9	21.6	7.6	5.8	470 ± 160	4.8 ± 5.2	3.7 ± 1.1

Source: Adapted from Hashimoto et al. (2021).

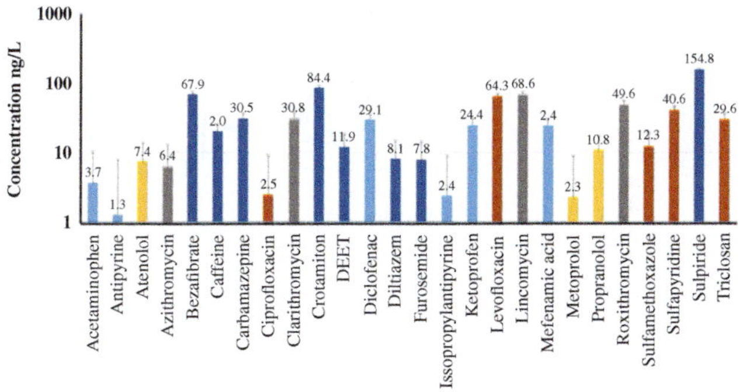

(a)

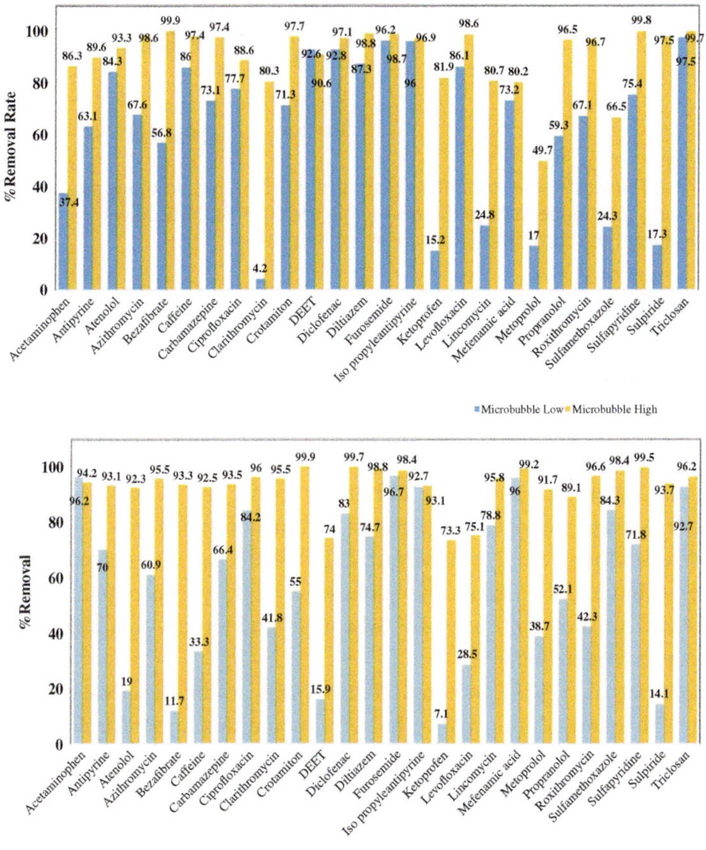

(b)

Figure 5.6 (a) List of pharmaceutical and personal care products found in secondary wastewater. and (b) the effect of O_3-MBs on the %removal of the Pharmaceutical and personal care products (PPCPs) by two types of MB generators. Source: Kim and Lee (2022) / Korean Society of Environmental Engineers.

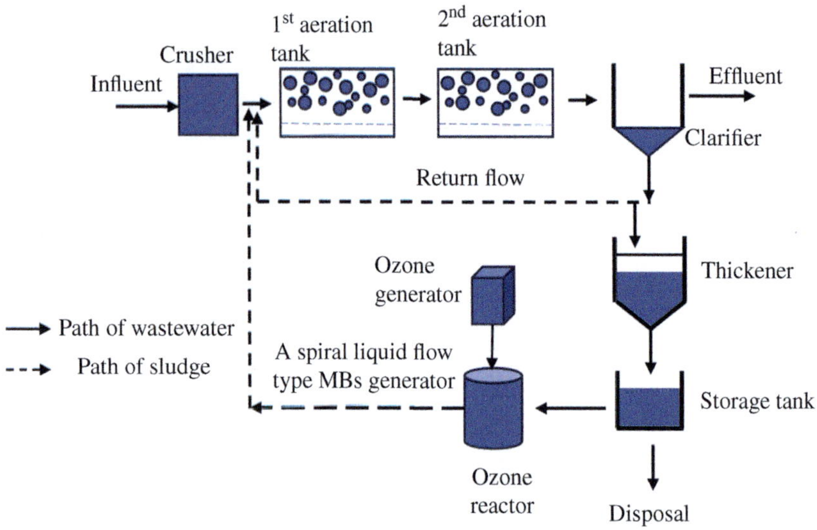

Figure 5.7 The wastewater treatment plant proposed by Hashimoto et al. (2021) for improving the sludge removal process using the O_3-MBs. (reused with permission from the copyright holder, Elsevier Publishing).

5.4 Tertiary Treatment

5.4.1 Disinfection

The MBs are extensively applied in the disinfection of water, mostly using O_3 as the gaseous phase. Bacterial contamination and subsequent infections are recognized as being a major threat to human health, and there is an urgent need to inactivate pathogenic organisms and prevent the spread of waterborne diseases to users and the environment. In this regard, the development of novel technologies based on the application of O_3-MBs is of paramount importance. A list of applications of MBs in several tertiary treatment processes is provided in Table 5.5.

Furuichi et al. (2013) reported that O_3-MB water deactivates both Gram-positive and Gram-negative bacteria, while this approach does not show any cytotoxicity against human gingival fibroblasts, unlike conventional mouthwash. Dissolved ozone concentration of 1.5 mg/L provided sufficient bactericidal activity for periodontal pathogens. Specifically, the inactivation of the bacterial cells (*Staphylococcus aureus*-2.4 × 108 CFU/mL, *Streptococcus sanguinis*-1.5 × 108, *Klebsiella pneumoniae*-7.6 × 10^8 CFU/mL, and *Escherichia coli*-1.6 × 109 CFU/mL) was >99.99% since the viable bacteria were below the detection limit (<10 CFU/mL). For *Porphyromonas*

Table 5.5 MB application in the tertiary treatment process (work in 2010–2024).

Advanced oxidation process (AOP)								
Wastewater type	**MB size**	**Gas phase**	**Flow rate**	**Temperature**	**pH**	**BOD, COD, TOC value**	**Scale**	**Reference**
Synthetic	1–25 μm	O_3	0.30–0.60 m/L	10–30	4–10		Lab Scale	Lee et al. (2019)
Industrial wastewater	40–370 nm	O_3	1 L/min	20 C	3–11	63%	Lab Scale	Xia and Hu (2019)
Synthetic wastewater	146.9 ± 19.7 nm			25C	3–12		Lab scale	Ma et al. (2023)
Synthetic wastewater	50 μm		0.3 L/min	25C	4–8		Lab Scale	Liu et al. (2018)
Industrial wastewater	10–50 μm	O_3	50 mg/L	25 C	7.0–7.5	88%	Lab Scale	Jothinathan et al. (2021)
Industrial wastewater	MNB	O_3	0.4 L/min	26C	7	43%	Lab scale	Adirajasa et al. (2022)
Synthetic wastewater	MB	O_3	1.1 mg/S		4–9		Pilot plant	Khuntia et al. (2014)
Synthetic wastewater	25 μm	O_3	8–80 ml/sec	25	4–10	80%	Pilot plant	Khuntia et al. (2014)
Raw secondary effluent	30–60 μm	O_3	1.2 Lpm	13–15 C		Total OD: 7.4–10.5 mg/L	Lab scale	Sumikura et al. (2007)
Acrylic fiber manu. Unit	<45 μm	O_3	48 mg/L	20 C	8	BOD/COD: improved from 0.18 to 0.47	Lab scale	Zheng et al. (2015)
Sludge removal in tertiary treatment step								
Coal-industry water		O_3	2.0 L/min	26.7C	8.8	60.882%	Lab Scale	Liu et al. (2018)
Disinfection								
Salmonella	<100	air		37	7		Laboratory	Yi et al. (2022)
Escherichia coli	<100	O_3	2.0 L/min	28	6.0–6.5		Laboratory	Saijai et al. (2019)

gingivalis cells with an initial bacterial concentration of 7.0×10^7 CFU/mL, the percentage of killed bacteria was higher than 99.99%, while the disinfection activity deteriorated in the case of *Streptococcus mutans* with an initial bacterial concentration of 1.7×10^6 CFU/mL since it reached maximum disinfection of 94.69% within three minutes. Another study for the evaluation of the bactericidal activity against periodontal pathogenic bacteria (*P. gingivalis* and *Aggregatibacter actinomycetemcomitans*) reported that O_3-NBs water with a concentration of 1.5 mg/L was capable of reducing the numbers of colony-forming units (CFU/mL) below the detection limit (<10 CFU/mL) after only 0.5 min of exposure, providing evidence that it is not cytotoxic to cells of human oral tissues.

5.4.2 Dyes Wastewater Treatment

The lab-scale study of the application of MBs for the degradation of dyes in synthetic and real wastewater is commonly found in the literature (Khuntia et al., 2014; Xia and Hu, 2019). The major factors affecting the efficiency of the %dye removal are the O_3 flow rate and O_3 concentration in the water phase. The dyes can decolorize rapidly by the O_3, whereas MBs help in the higher rate of mass transfer of O_3 in water, as shown in Figure 5.8. The decolorization of the dyes is very fast due to the breakage of the conjugated bond by the strong oxidation potential of O_3. The conjugated bonds are mostly responsible for the color of the water-soluble dyes, as shown in Figure 5.9. However, the complete mineralization of the dye compound along with the fragmented intermediates due to the attack of O_3 might take much longer. For example, the commonly used Brilliant green (BG) dye, $C_{23}H_{26}N_2O$, at an initial concentration of 0.0274–0.164 mM (50–250 mg/L), can take 30 min for decolorization at O_3 generation rate of 1.7 mg/s in a pilot-scale MB system of 20 L capacity. However, without MBs, the dye can take up to one hour to be colorless at the above capacity of O_3 generation and water volume. However, upon complete analysis of the degradation process and pathway, it was found that many intermediates were formed during the degradation of the dye (as shown in Figure 5.10), and to reach the final stage of mineralization, the reaction takes up to one hour.

5.5 Ozonation and Biological Process

The combination of O_3 and biological processes for the treatment of water and wastewater is an effective method to treat in the tertiary step. A typical schematic of the laboratory scale of the combination of MBs and bioreactor

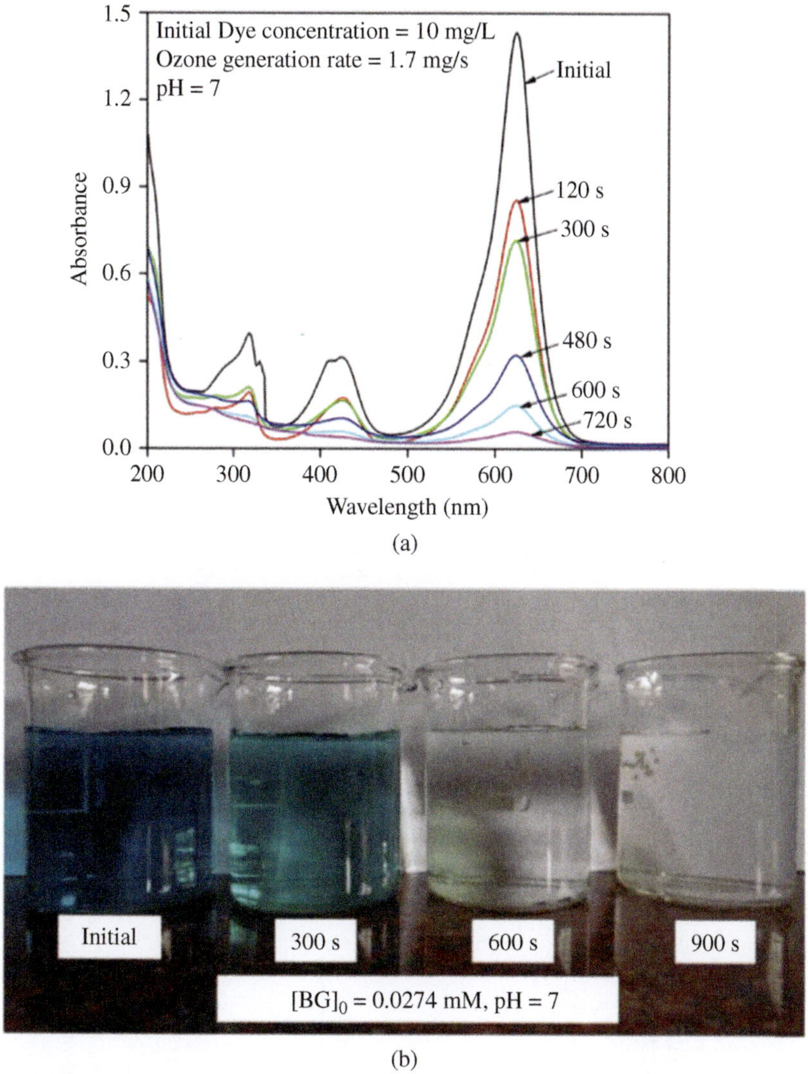

Figure 5.8 The application of O_3-MBs for the treatment of a synthetic dye wastewater (Brilliant Green, BG), (a) the UV–Vis spectrometer of the dye samples during the O_3-MB treatment process, (b) visualization of variation of BG samples treated with O_3-MB with respect to time. Source: Reproduced with permission from Khuntia et al. (2014)/ELSEVIER.

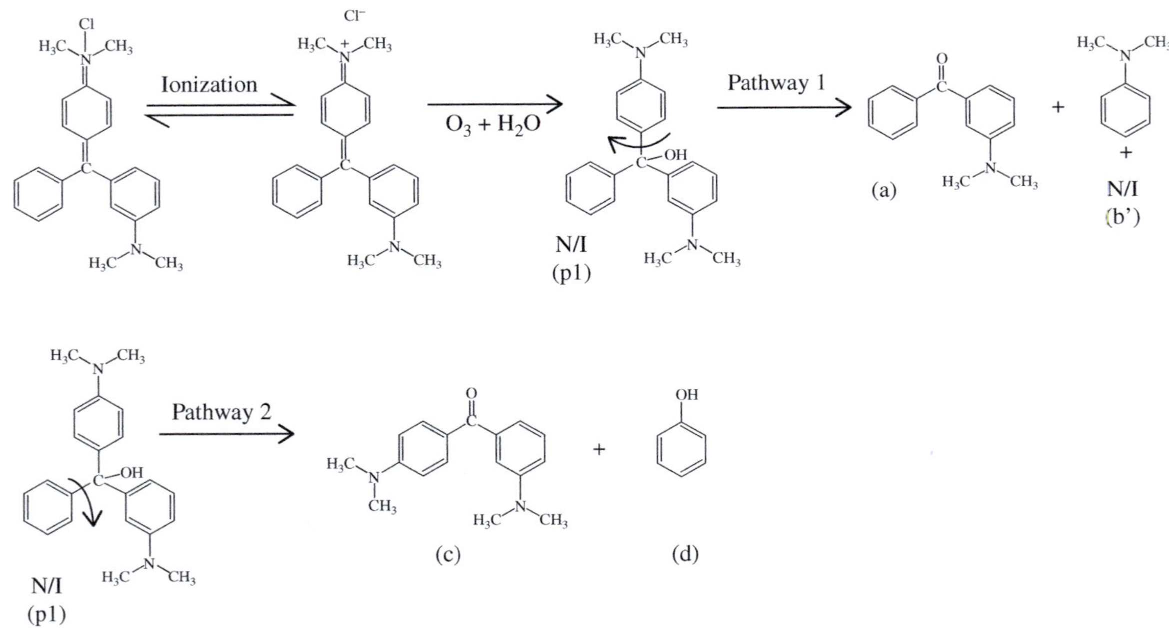

Figure 5.9 The ionization and formation pathways of intermediates during the degradation of BG dye in the ozonation process. Source: Khuntia et al. (2014)/with permission of ELSEVIER.

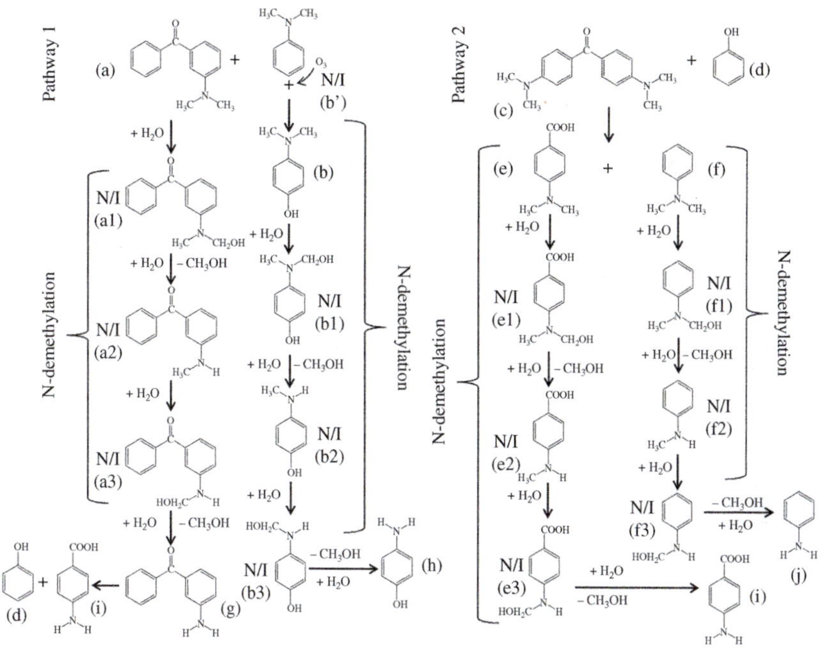

Figure 5.10 The detailed degradation pathway of degradation of BG dye in the O_3-MBs process and formation of several intermediates. Source: Khuntia et al. (2014)/with permission of ELSEVIER.

is shown in Figure 5.11. The MBs reactor is a 25 L capacity sealed pressure vessel. It was a one-of-a-kind reactor designed to treat bio-treated coal-based wastewater. Microbubble-catalytic-O_3-reactor (MOCR) and biological reactors are connected in series, and O_3 gases are used in the gas phase of MBs. The MCOR has three layers of 8–10 mm of coal-based activated carbon as the catalyst (Liu et al., 2018).

5.6 Fiber, Pharmaceuticals, and Other Wastewaters

To enhance the mass transfer as well as the generation of free radicals, many combinations of techniques have been used. Since MBs are able to enhance the mass transfer of the powerful O_3, other aids such as UV, homogeneous and heterogeneous catalysts, and ultrasonication processes are combined with O_3 for the enhanced generation of various free radicals (Lee et al., 2019). Most of the integrated methodologies are applied for the degradation of benign and persistent micropollutants found in water, such as pharmaceuticals and fibers, whose concentration at the level of mg/L to

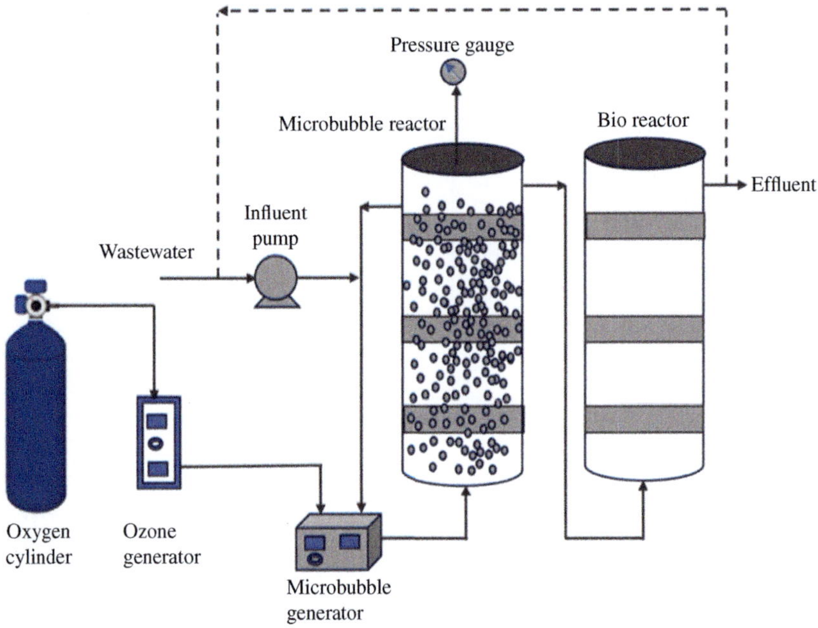

Figure 5.11 The integration of ozonation and biological processes for the treatment of tertiary wastewater. Source: Liu et al. (2018)/with permission of ELSEVIER.

μg/L can also be dangerous for human health (Zheng et al., 2015). Therefore, synergistic improvisation has been followed for these effluents so that the chance of release of trace amounts of these compounds can be avoided. One such combination of UV irradiation with O_3-MBs has been designed by Zheng et al. (2015), as shown in Figure 5.12, a setup designed for the treatment of manufacturing wastewater from the acrylic fiber industry. The reaction column comprises the UV lamp and the MBS column in a single unit where the UV lamp is situated in the core, considering that the MBs can be activated to generate radicals directly and faster in this setup while the MBs rise in the water.

5.7 Integration of Microbubble with Membrane Process for Water Treatment

Ever since membrane technology was invented, it has been the most exploited method for the tertiary step of drinking water treatment. Polymeric membranes are mostly used for the resultant micropollutant removal

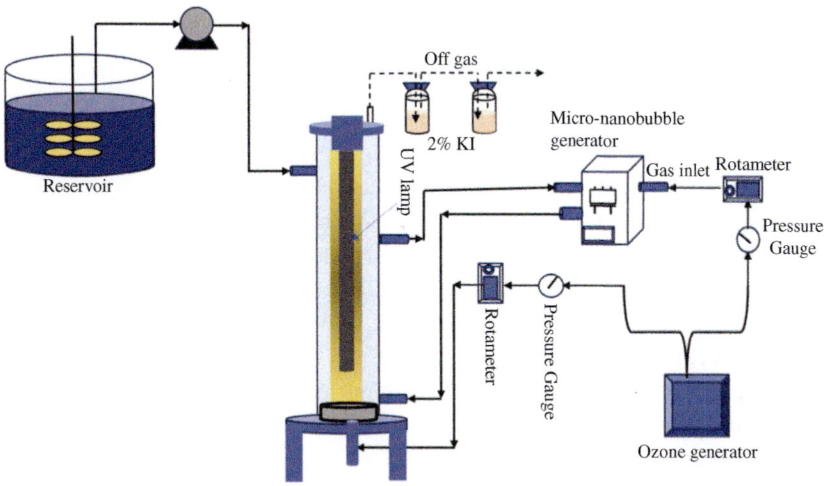

Figure 5.12 A single setup for the O_3-MB-UV-based bubble column. Source: Adapted from Zheng et al. (2015).

and can be designed for any scale of operation. However, the persistent problem in the membrane separation is membrane fouling, which then makes the process energy- and cost-intensive while operated on an industrial scale. The MBs have shown impressive efficiency in the membrane de-fouling process making the membrane life much longer (Ye et al., 2019; Wu et al., 2008). The traditional methods for membrane de-fouling include backward washing, gas sparing, and chemical cleaning. The application of MBs in membrane de-fouling is a new technique that has a larger scope of application for industrial-scale membrane technology applications. However, the usage of MBs integrated with membrane technology is aimed at membrane cleaning, where MBs are not directly involved in water treatment. Hence, more about the integration of membranes and MBs, along with other contemporary applications of MBs, have been discussed in Chapter 6.

References

Adirajasa, Y., Ahmad, F.A., Eko, Y. et al. (2022). Study of application advanced oxidation process (AOP) using nano-micro bubble ozone for batik industry. Waste Water Treatment. IJISRT 7: 800–809.

Agarwal, A., Ng, W.J., and Liu, Y. (2011). Principle and applications of microbubble and nanobubble technology for water treatment. Chemosphere 84: 1175–1180.

Ali, F., Azmi, K.N., and Firdaus, M.R. (2021). The effectiveness of microbubble technology in the quality improvement of raw water sample. Mater. Sci. Eng. 1144: 012053.

Bai, Z., Wang, J., and Yang, Q. (2017). Advanced treatment of municipal secondary effluent by catalytic ozonation using Fe_3O_4-CeO_2/MWCNTs as an efficient catalyst. Environ. Sci. Pollut. Res. 24: 9337–9349.

Chua, B.L., Xing, X.H., Yu, F.A. et al. (2008). Enhanced treatment of practical textile wastewater by microbubble ozonation. PSEP. 86: 389–393.

Fan, M., Tao, D., Honaker, R., and Luo, Z. (2010). Nanobubble generation and its application in froth flotation. J. Min. Sci. 20: 01–19.

Fan, W., Zhou, Z., Wang, W., and Huo, M. (2019). Environmentally friendly approach for advanced treatment of municipal secondary effluent by integration of micro-nano bubbles and photocatalysis. J. Clean. Prod. 237: 117828.

Farajzadehha, S., Saghravani, F.S., and Shayegan, J. (2022). Pilot-scale study of hydraulic retention time and energy consumption in biological treatment of raw municipal wastewater by air micro-nano bubble aeration in different seasons. Desalin. Water Treat. 280: 9–18.

Furuichi, A., Arakawa, S., Mano, Y. et al. (2013). Comparative analysis of efficacy of ozone nano bubble water (NBW3) with established antimicrobials bactericidal efficacy and cellular response an in vitro study. J. Oral Tissue Eng. 10: 131–141.

Hashimoto, K., Nakai, S., Motoshige, H., and Nishijima, W. (2021). Sludge reduction in a full-scale wastewater treatment plant using ultra-fine-and micro-bubble ozonation. Sci. Eng. 43: 127–135.

John, A.A., Brookes, A., Irene, C.I. et al. (2020). Microbubbles and their application to ozonation in water treatment: a critical review exploring their benefit and future application. Environ. Sci. Technol. 52: 1561–1603.

Jothinathan, L., Cai, Q.Q., Ong, S.L., and Hu, J.Y. (2021). Organics removal in high strength petrochemical wastewater with combined microbubble-catalytic ozonation process. Chemosphere 263: 127980.

Kaushik, G. and Chel, A. (2014). Microbubble technology: emerging field for water treatment. Bubble Sci. Eng. Technol. 5: 33–38.

Khuntia, S., Majumdar, S.K., and Ghosh, P. (2012). Microbubble-aided water and wastewater purification: a review. Rev. Chem. Eng. 28: 191–221.

Khuntia, S., Majumder, S.K., and Ghosh, P. (2014). Oxidation of as (III) to as (V) using ozone microbubbles. Chemosphere 97: 120–124.

Kim, I. and Lee, J. (2022). Comparison of ozonation removal for PPCPs in secondary treated sewage by microbubble generator and ejector. Environ. Eng. Res. 27: 200163.

Lee, Y.G., Park, Y., Lee, G. et al. (2019). Enhanced degradation of pharmaceutical compounds by a microbubble ozonation process: effects of temperature, pH, and humic acids. Energies 12: 4373.

Levitsky, I. and Tavor, D. (2022). Vitaly Gitis. Micro and nanobubbles in water and wastewater treatment: a state-of-the-art review. J. Water Proc. Eng. 47: 102688.

Liu, C., Chen, X.X., Zhang, J. et al. (2018). Advanced treatment of bio-treated coal chemical wastewater by a novel combination of microbubble catalytic ozonation and biological process. Sep. Purif. Technol. 197: 295–301.

Liu, S., Wang, Q., Ma, H. et al. (2010). Effect of micro-bubbles on coagulation flotation process of dyeing wastewater. Sep. Purif. Technol. 71: 337–346.

Lucasa, M.S., Peresa, J.A., Bing, L.Y., and Li, P.G. (2009). Ozonation kinetics of winery wastewater in a pilot-scale bubble column reactor. Water Res. 43: 1523–1532.

Ma, P., Han, C., He, Q. et al. (2023). Oxidation of Congo red by Fenton coupled with micro and nanobubbles. Environ. Technol. 44: 2539–2548.

Menendeza, D. and Floresa, J.V. (2017). Reduction of hospital wastewater through micro-nano ozone-air bubbles. J. Nanotechnol. 1: 2522–6908.

Nair, S.S., Ruben, P.C., Stubbs, T. et al. (2022). Contemporary application of microbubble technology in water treatment. Water Sci. Technol 86: 2138.

Pal, P., Joshi, A., and Anantharaman, H. (2022). Nanobubble ozonation for waterbody rejuvenation at different locations in India: a holistic and sustainable approach. RINENG 16: 100725.

Ryskie, S., Carolina, G.M., Carmen, M.N., and Genty, T. (2020). Efficiency of ozone microbubbles for ammonia removal from mine effluents. Miner. Eng. 145: 106071.

Saijai, S., Thonglek, V., and K. Yoshikawa K. (2019). Sterilization effects of ozone fine (micro/nano) bubble water. IJPEST 12: 55–58.

Salgueroa, J. and Flores, J.V. (2017). Reduction of the biochemical oxygen demand of the water samples from the lower basin of the Chillon river by means of air-ozone micronanobubbles, Ventanilla – Callao. J. Nanotechnol. 1: 2522–6908.

Selihin, N.M. and Tay, M.G. (2021). A review on future wastewater treatment technologies: micro-nanobubbles, hybrid electro-Fenton processes, photocatalytic fuel cells, and microbial fuel cells. Water Sci. Technol. 85: 319.

Singh A., Dalvi S., Sinha M.K., Khuntia S., Magnetite nanoparticle mediated ozonation of ciprofloxacin wastewater: energy calculation and optimization of parameters. Chem. Eng. Commun. 2024. doi. https://doi.org/10.1080/00986445.2024.2362815

Singh, A., Sinha, M.K., and Khuntia, S. (2022). Enhanced generation of hydroxyl radicals by OMS-2 catalyst for flue gas absorption. Mater. Today. Proc. 57: 279–284.

Sumikura, M., Hidaka, M., Murakami, H. et al. (2007). Ozone micro-bubble disinfection method for wastewater reuse system. Water Sci. Technol. 56: 53–61.

Suwartha, N., Syamzida, D., Priadi, C.R. et al. (2020). Effect of size variation on microbubble mass transfer coefficient in flotation and aeration processes. Heliyon 6: e03748.

Tekile, A., Kim, I., and Lee, J.Y. (2017). Applications of ozone micro- and nanobubble technologies in water and wastewater treatment. KSWW 31: 481–490.

Wu, Z.H., Chen, H.B., Dong, Y.M. et al. (2008). Cleaning using nanobubbles: defouling by electrochemical generation of bubbles. J. Colloid Interface Sci. 328: 10–14.

Xia, Z. and Hu, L. (2019). Treatment of organics contaminated wastewater by ozone micro-nano-bubbles. Water 11: 55.

Yao, K., Chi, Y., Wang, F. et al. (2016). The effect of microbubbles on gas-liquid mass transfer coefficient and degradation rate of COD in wastewater treatment. Water Sci. Technol. 73: 1969–1977.

Ye, Y., Yu, S., Hou, L. et al. (2019). Microbubble aeration enhances performance of vacuum membrane distillation desalination by alleviating membrane scaling. Water Res. 149: 588–595.

Yi, S.H., Jung, Y.H., Oh, S.I. et al. (2022). Synergistic antibacterial effect of disinfectants and microbubble water to salmonella Typhimurium. Korean J. Vet. Serv. 45: 277–284.

Zheng, T., Zhang, T., Wang, Q. et al. (2015). Advanced treatment of acrylic fiber manufacturing wastewater with a combined microbubble-ozonation/ultraviolet irradiation process. RSC Adv. 5: 77601–77609.

Zhou, S., Marcelino, K.R., Wongkiew, S. et al. (2022). Untapped potential: applying microbubble and nanobubble technology in water and wastewater treatment and ecological restoration. ACS EST Engg. 2: 1558–1573.

6

Other Industrial Applications of Microbubbles

6.1 Introduction

Up until now, a lot has been discussed on the noteworthy properties of microbubbles (MBs) in the context of hydrodynamics, mass transfer, and ability to enhance the generation of reactive oxygen species or radicals. However, another aspect of the MBs is their ability to enhance the flotation of micro-sized particles, which makes the MBs more versatile and contemporary in application outside of the water treatment domain. In this chapter, some more nonmedical, but industrial applications of MBs will be discussed, which shows the potential of MBs to outsmart the conventional techniques. At the same time, the MBs have been proven to be a more sustainable method of application. Similar to the water micropollutants, the other microscopic particles in nature are also difficult to handle. In mining and mineral beneficiation, the problem of micro- or ultrafine particles is a matter of concern that are difficult to separate from the air or water. Other than that, MBs can be used to enhance the performance of cleaning, scrubbing, biodiesel production, and reactive distillation. This chapter is dedicated to discussing the role of MBs in these important industrial processes.

6.2 Flotation and Mineral Beneficiation

The conventional method for the separation of fine and ultrafine particles from a solid–solid mixture, or solid–liquid mixture, is flotation, or froth flotation. While working with MBs for mineral flotation, it is essential to use or avoid using additional surfactants and foaming agents used in the froth flotation processes and fully utilize the slow-rise speed of the MBs to carry the lighter particles to the water surface. In the conventional macrobubbles, the size of the bubbles is too high, and hence the rising velocity is also

Microbubbles: Engineering Aspects and Industrial Applications, First Edition. Snigdha Khuntia.

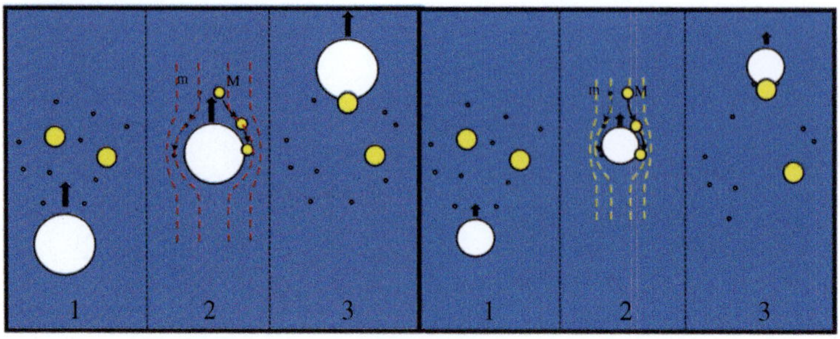

Figure 6.1 Rise of bubbles (a) in a conventional bubble column for macrobubbles and (b) MBs colloid with the hydrobhobic particles (black indicates fine particles, yellow indicates course particles, white indicates the bubbles. Source: Jung et al. (2023)/with permission of ELSEVIER.

higher. Therefore, the collision between the particles and bubbles is too less due to shorter residence time. To enhance the attraction of the bubbles and maintain the bubble size uniformity, surface active agents are used.

The use of MBs is not a new process and has been studied by many researchers for the mineral separation in ultrafine size range of up to 100 μm. (Ahmed and Jameson, 1985; Etchepare et al., 2017; Gontijo et al., 2007; Oliveira et al., 2018; Ralston et al., 1999).

The mechanism of attachment of minerals to the surface of MBs followed by rise to the surface for separation is shown in Figure 6.1. The fine particles will have a lower inertia, and the MBs, due to their higher rise velocity, will have a higher drag force. Hence, instead of attaching to the bubble surface, the particles will surpass the movement of the bubble to the surface and miss the chance of getting attached. On the contrary, the MBs will have lower drag force, and the fine particles will likely get enough time to attach to the bubble surface and will remain attached till they reach the top. The MBs surface charge also plays an important role in the attachment process of the minerals, which can be tuned using a small amount of surface active groups.

In flotation process, a generic polydispersity index (PI) is used to evaluate the scope application of the bubbles, which is correlated to the bubble size. Unlike biomedical applications, the industrial process like flotation requires MBs of more than 10 μm. A precise comparison of PI applicable for flotation and water treatment is shown in Figure 6.2.

The term PI is defined as:

$$\mathrm{PI} = \frac{d_{90} - d_{10}}{2 \times d_{50}} \tag{6.1}$$

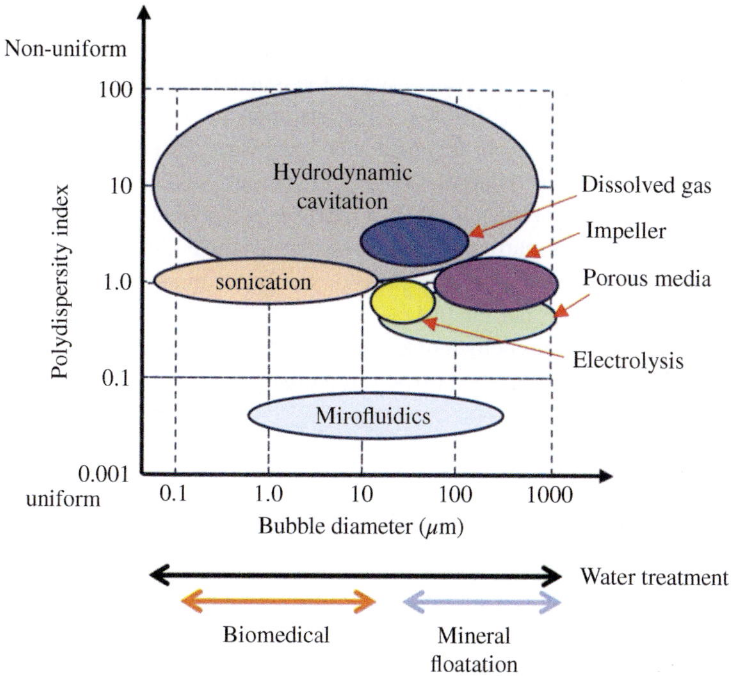

Figure 6.2 The profile of polydispersity index (PI) and bubble diameters used for various processes, where PI is the indicative of bubble uniformity. Source: Jung et al. (2023)/ with permission of ELSEVIER.

where the subscripts 10, 50, and 90 represent the 10%, 50%, and 90%, respectively, of the cumulative particle size in the size distribution curve. The figure also provides ample information on the type of MB-generating devices preferred for the application.

In a nutshell, the MB-based flotation of minerals follows three major steps for separation:

(i) MBs aeration: Introduction of MBs in the flotation column

(ii) Mixing of MBs and fine/ultrafine particles to allow particle–bubble collision

(iii) Separation of MBs and particles at the surface followed by skimming of particles formed in layers.

6.2.1 Theory of Flotation

The flotation process to carry the particles with MBs to the surface of water can be theoretically expressed by Newton's formula.

$$mdv = F_B + F_D + F_G \tag{6.2}$$

where F_B, F_D, and F_G represent the forces acting on the particle and the bubble, such as buoyancy force, drag force, and gravitational force, respectively. The buoyancy force F_B acts upward, while the gravitational force F_G pulls the bubble downward. The drag force F_D acts opposite to the direction of motion of the body, which in this case acts downward. Initially, when the acceleration of the flocs of particles is zero, the amount of dirt is also small, and the overall shape of the particle-laden MBs is considered as a spherical ball. Using the general mathematical expressions for the F_B, F_D, and F_G and doing an overall force balance, an expression for the terminal velocity of the bubbles is calculated as per Eq. (3.1).

The C_D or bubble drag coefficient is considered as 0.4, and the rise velocity u_p for a particle rising with the bubble is modified as:

$$u_p = 1.83\sqrt{gd_p\frac{\rho_L - \rho_p}{\rho_L}} \tag{6.3}$$

Under the extremely high laminar flow regime, the dirt particle will flow according to Stoke's law and attain a velocity as:

$$u_p = \frac{g}{18\mu_p}(\rho_L - \rho_p)d_p^2 \tag{6.4}$$

For the transition flow regime, the expression for u_p for a particle rising with the bubble is modified as:

$$u_p = 0.22\left(\frac{\rho_L - \rho_p}{\rho_L}\right)^{2/3}\frac{d_p}{v_p^{1/3}} \tag{6.5}$$

From Eqs. (6.2)–(6.5), it becomes evident that the rising velocity of the dirt-laden MBs when moving toward the water surface is largely influenced by three key factors: (1) the difference in density between the dirt and water; (2) the fluid flow regime under which the flow occurs, such as laminar, turbulent, or transition; and (3) the size of the MBs. Notably, having the smaller bubble diameters enables the adherence of more bubbles to a unit volume of dirt, reducing the overall density of the dirt–bubble mixture and consequently enhancing the flotation speed. Accelerated flotation velocity corresponds to shorter water separation times and increased separation efficiency, underscoring a crucial advantage of air flotation over traditional sedimentation methods.

6.2.2 Kinetics of Microbubble (MB)-Induced Flotation

The kinetics of particles and bubbles have been studied in the context of the hydrodynamic behavior of the particle–bubble mixture. Bloom and Heindel (1997) have modeled the first-order rate kinetics of the particle–bubble mixture to predict the rate of removal of particles.

$$\frac{dN_P}{dt} = -K_F N_P \tag{6.6}$$

where N_P represents the number density of fine minerals that are floatable and ready for attachment to the bubble surface and K_F is the rate constant for the particle–MBs flotation. To obtain a larger value of K_F, the flotation process is also dependent on MBs and mineral particle size, collision frequency (Z), and particle–MBs hydrodynamics. To be precise, Eq. (6.6) is further modified as

$$\frac{dN_P}{dt} = ZE_C E_a E_s \tag{6.7}$$

The particle collision efficiency E_c, adhesion efficiency E_a, and stability efficiency E_s are equally responsible for the rate constant for the particle–MBs flotation. The value of Z can be measured by the empirical equation as (Pyke et al., 2003),

$$Z = 5D_{MB} N_P \left(\frac{d_b + d_F}{2}\right)^2 \sqrt{(V_{MBs}^2 + V_F^2)} \tag{6.8}$$

D_{MB} is the MB density in the column, calculated as shown in Eq. (6.9). The root mean square velocity is used for both MBs and fine particles, denoted as V_{MBs} and V_F, respectively.

$$D_{MB} = \frac{6\epsilon_G}{\pi d_b^3} \tag{6.9}$$

The root mean square velocity V_{MBs}, or rms of MBs, is expressed as

$$\sqrt{V_{MBs}^2} = \frac{0.33\varepsilon^{4/9} d_b^{7/9}}{\mu_l{}^{1/3}} \left(\frac{\Delta\rho}{\rho_F}\right)^{2/3} \tag{6.10}$$

ε represents the rate of energy dissipation per unit mass, ν is the fluid kinematic viscosity, and $\Delta\rho$ is the difference in the density of fine particle ρ_F and the fluid.

An expression to evaluate the ε has been suggested by Wu and Patterson (1989) and Pope (2000), respectively.

$$\varepsilon = 0.85 \frac{\left[\frac{U_{c,x}^2 + U_{c,y}^2 + U_{c,z}^2}{2}\right]^{3/2}}{L} \tag{6.11}$$

$$\varepsilon = 0.85 \frac{\left[\frac{U_{c,x}^2 + U_{c,y}^2 + U_{c,z}^2}{2}\right]^{3/2}}{D_C} \tag{6.12}$$

$U_{c,x}$, $U_{c,y}$, and $U_{c,z}$ represent the particle–MBs mixture circulation velocity in x, y, and z axis, and L is the flow length scale, which can be considered as 0.07 D_c, as suggested by Pope (2000).

Incorporating the expressions of V_{MBs}, ε, and Z in Eq. (6.6) and simplifying further, K_F is expressed as:

$$K_F = 30\frac{\epsilon_G}{\pi D_{MBs}^3}\left(\frac{d_b + d_F}{2}\right)^2\left(\frac{d_b^{7/9}}{\nu^{1/3}}\right)\left(\frac{\Delta\rho}{\rho_F}\right)^{2/3}\left[\frac{\left(\frac{U_c^2}{2}\right)^{3/2}}{D_c}\right]^{4/9} E_C E_a E_s \tag{6.13}$$

The % recovery of the fine minerals (R_c) after the flotation process at time t is a function of K_F and evaluated by:

$$R_c = R_{max}(1 - \exp(-K_F t)) \tag{6.14}$$

R_{max} is the maximum possible recovery. The particle collision efficiency E_c, adhesion efficiency E_a, and stability efficiency E_s are evaluated through many correlations available in the literature, which involve the concepts of fluid dynamics and particle fluid hydrodynamics. To explain much about these is out of the context of this book, but one can refer to Weber and Paddock (1983), Dobby and Finch (1987), and Parmar and Majumder (2016).

The addition of surface active agents such as sodium dodecyl sulfate (SDS) and cetyltrimethylammonium bromide (CTAB) can surely enhance the K_F, and when measured against *Uc*, follows a trend shown in Figure 6.3.

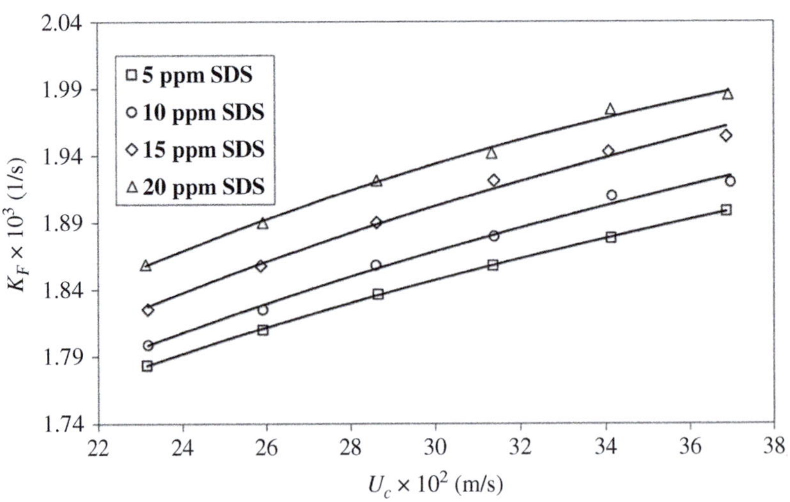

Figure 6.3 Effect of circulation velocity on the flotation rate constant of fine CuO particles in the presence of different surfactant-mediated MBs. Source: Parmar and Majumder (2016)/with permission of ELSEVIER.

A lot of lab- and industrial-scale studies have been reported on the flotation of precious fine minerals and particles from dilute solutions, such as separation of Molybdenum from Tungsten (Su et al., 2022), quartz and apatitic (Rosa and Rubio, 2018), magnesite (Tsave et al., 2021), P_2O_5-apatite (Liu et al., 2020), and many more. The separation efficiency of MB system in order to separate the fine and ultrafine minerals is calculated on the basis of several indices.

$$\eta_M = \left(1 - \frac{C_M}{C_F}\right) \times 100\% \tag{6.15}$$

η_M is the mineralization efficiency, and C_M and C_F are the concentrations of target mineral/s in supernatant and feed, respectively.

$$\eta_F = \left(1 - \frac{C_R \times V_R}{C_F \times V_F}\right) \times 100\% \tag{6.16}$$

η_F is the flotation efficiency, C_R is the concentration of target mineral/s in raffinate, and V_R and V_F are the volumes of raffinate and feed, respectively. While calculating the MB-aided scrubbing efficiency of the minerals, the calculation involves a separate set of parameters related to the total mass balance of the scrubber. The scrubbing efficiency (η_{SC}) is calculated as:

$$\eta_{SC} = \frac{C_{SC} \times V_{SC}}{C_{LO} \times V_{LO}} \times 100\% \tag{6.17}$$

C_{SC} and C_{LO} are the concentrations of target mineral/s in scrubbing liquor and floated organic phase, respectively. V_{SC} and V_{LO} are volumes of scrubbing liquor and floated organic phase, respectively. Similarly, the stripping efficiency (η_{ST}) is calculated as:

$$\eta_{ST} = \frac{C_{ST} \times V_{ST}}{C_{LO} \times V_{LO}} \times 100\% \tag{6.18}$$

C_{ST} is the concentration of target mineral/s in stripping liquor, and V_{ST} is the volume of stripping liquor. Some other indices, such as distribution coefficient (D_{Coeff}), are also commonly used in the mineral flotation experiments, defined as:

$$D_{Coeff} = \frac{C_{LO}}{C_R} \times 100\% \tag{6.19}$$

6.3 Cleaning and Degreasing

Metallic components and parts are used in assembling several electronic components and devices. The market demand and supply of electronic devices are rising each year, and the requirement for the metallic parts for

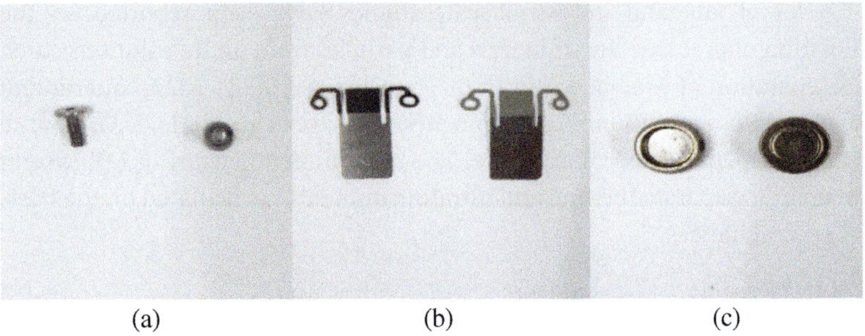

Figure 6.4 Various small metallic parts to clean with MBs cleaning method. Source: Tan et al. (2020)/ with permission of ELSEVIER.

these devices is up to a billion pieces a year. During the fabrication, cutting, and molding of metallic parts, a lot of heat is generated. Along with that, proper cleaning of metallic parts is also important before assembling them in the device; otherwise, the contaminated metals can cause several glitches and damage to the device. More caution is required for the fabrication of laboratory and medical devices, and hence any parts, irrespective of their size, have to be of the utmost cleanliness and free from any dirt deposition.

Most industries are dependent on purified water for cleaning metal plates and parts. Whereas some advanced methods are ultrasound, vapor degreasing, solvent/surfactant cleaning, and mechanical and immersion cleaning. Ultrasonic cleaning is a rapid and effective method, but at the same time could be risky for the transducer if it remains unattended. However, ultrasonic processes are heavily power-consuming and pose a high risk of noise pollution.

The solvent cleaning process is another method where the commonly used solvents are carbon tetrachloride (CCl_4), tetrachloroethylene, or trichloroethylene is used with water for the cleaning of the metallic parts. However, these solvents are responsible for the ozone depletion and are known to have a high risk of health and safety. Instead of these, some common surfactants are used for the metal cleaning, such as SDS and CTAB, as shown in Figure 6.4 (Tan et al., 2020). However, these surfactants can also leave residual amounts and spots on the metal surface and shall be removed completely before assembling the device. The extra caution and thorough cleaning of the miniature metallic parts pose a surplus demand for clean water for the industry.

Bubble-based cleaning using MBs and nanobubbles (NBs) is a hassle-free method for cleaning any size of metallic parts where the parts are immersed in a simple cleaning clamber containing MBs and NBs, generated in situ.

Taking the advantage of their size, this tiny bubble can penetrate to the corners of any size and shape of the metallic parts and clean the entire surface. The parts as small as screws and caps can be cleaned without leaving any dirt spots. In the experimental evaluation, the measurement of grease/dirt content of the metal parts is expressed by weighing their mass, and a generic expression for the %grease/dirt removal is used to evaluate the efficiency of cleaning.

$$\%Grease\ removal = \frac{W_{metal,before} - W_{metal,after}}{W_{metal,before}} \times 100\% \tag{6.20}$$

Degreasing is another challenging and water-consuming process for the smaller metallic parts in the manufacturing industries. Earlier, chlorofluorocarbon (CFC) was used heavily for the degreasing of the metallic pieces. But since the use of CFC has been put under environmental surveillance, the only option was to use surfactant-loaded water for degreasing. While MBs were gathering attention in the water treatment and other processes, and many industrial MB generators were designed and fabricated, the degreasing of metal was also tried and tested, taking the leverage of the size and properties of the MBs. Miyamoto et al. (2007) have closely observed the intrinsic steps involved in the process of degreasing the metals through a high-speed camera, as shown in Figure 6.5, and have summarized them as follows:

Step 1: The hydrophilic grease/oil gets adsorbed to the outer surface of the tiny MBs filled with air.

Step 2: The grease/oil molecules cover the entire surface of the and try to stabilize the MBs by exerting a high energy state at the oil–water interface. The grease/oil acts like a stable coating on the bubble surface in this state.

Step 3: The MBs, due to their rising velocity, but not coated with the grease/oil, come to the water surface altogether without collapsing under water.

This whole process makes the detachment of the grease/oil from the metal surface easier without adding the surfactants. However, to clean heavy load of grease/oil, a small amount of additive (less than 0.01 wt%) is sometimes recommended to enhance the process of degreasing.

Factors such as temperature, pH, hydraulic speed, and degreasing time can affect the entire process of degreasing. However, the degreasing time is the most important parameter for the best result, where proving sufficient time for the attachment of the grease/oil molecules to the MBs can give a better result. Since the MBs can form hotspots and lead to corrosion on metal plates, the temperature shall be maintained below 40 °C (Li et al., 2022). While performing the degreasing of the stainless steel metal plates

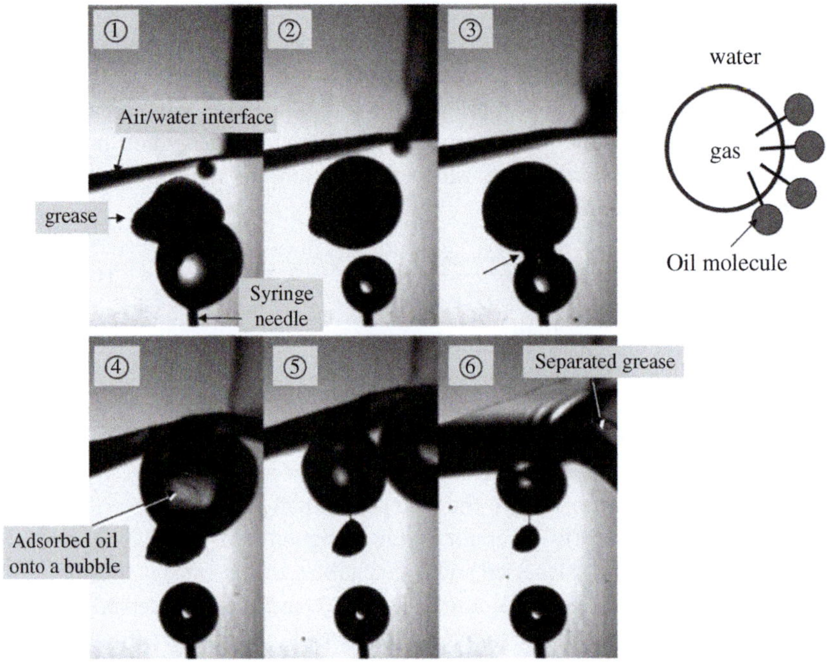

Figure 6.5 Photographs captured in the degreasing process, showing the absorption and detachment of oil attached to MBs. Source: Reproduced with permission from Miyamoto et al. (2007)/IOP Publishing.

with and without MBs, Li et al. (2022) performed a visual experiment under the fluorescent light source to detect visible effects of MBs (Figure 6.6). Another interesting work has been reported in the degreasing of metal plates using O_3-MBs (Ruoyi et al., 2020).

6.4 Scrubber

Gas absorption or scrubbing of harmful gases like SO_2 and NO_X using water as the scrubber is a known process for industries. Scrubbing the multiphase components like SO_2 and NO_X gases in water is a difficult task for multiple reasons. Out of several challenges in the scrubbing process, one is the poor solubility of NO_X in water, which requires special attention for scrubbing. The wet flue gas treatment processes not only look for solution to increase the solubility of NO_X in water but also intend to oxidize both SO_2 and NO_X in the water phase in a single column. Because scrubbing of the flue gases demands a lot of pure water and energy, a simultaneous oxidation of these gaseous compounds can save a fortune for the industries.

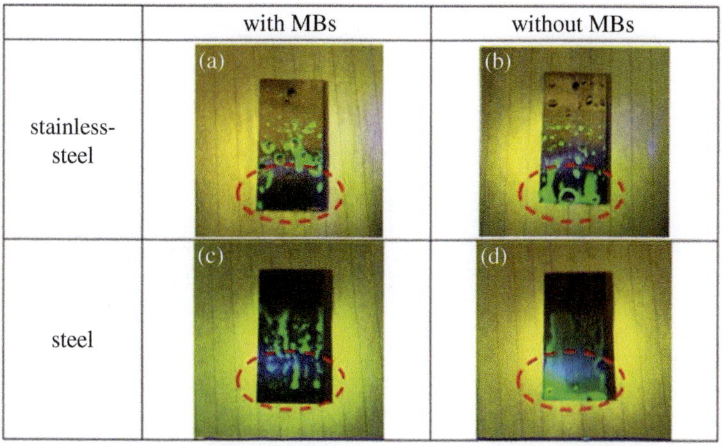

Figure 6.6 Degreasing of metal plates and steel pieces using MBs, captured by the fluorescence images; the green showing the oil attached to the plates decreased (a) stainless steel parts with MBs, (b) stainless steel parts without MBs, (c) plain steel plates with MBs, and (d) plain steel parts without MBs. Source: Reproduced with permission from Li et al., 2022/ELSEVIER.

A few efforts have been made to enhance the solubility of the NO_X using MBs. It is known that SO_2 is highly water-soluble and hence easier to remove (Xiao and Li, 2019, Xiao et al., 2020). But to treat both flue gases simultaneously, oxidizing agents such as H_2O_2 and O_3 can be used. O_3-MBs help in the generation of $OH^{\bullet}$ radicals, hence can react to oxidize the NO_X and SO_2 in water. In comparison to the other oxidation/reduction wet methods for simultaneous treatment of NO_X and SO_2, the use of MBs was promising to achieve near 100% removal.

In one interesting work reported by Yoo et al. (2023), they presented a contemporary application of MBs for extraction of some valuable products from seawater bittern using an experimental route, as shown in Figure 6.7. The MB scrubber was used to minimize air pollution simultaneously during the process of extractions, where the MBs were solely used for the treatment of gaseous pollutants such as SO_2 and NO_X, avoiding the addition of any other chemicals.

6.5 Biodiesel Production

Transesterification stands out as a top method for biodiesel production, even though certain feedstocks, like nonedible oils, carry higher levels of free fatty acids (FFAs). Traditional biodiesel production leans toward homogenous

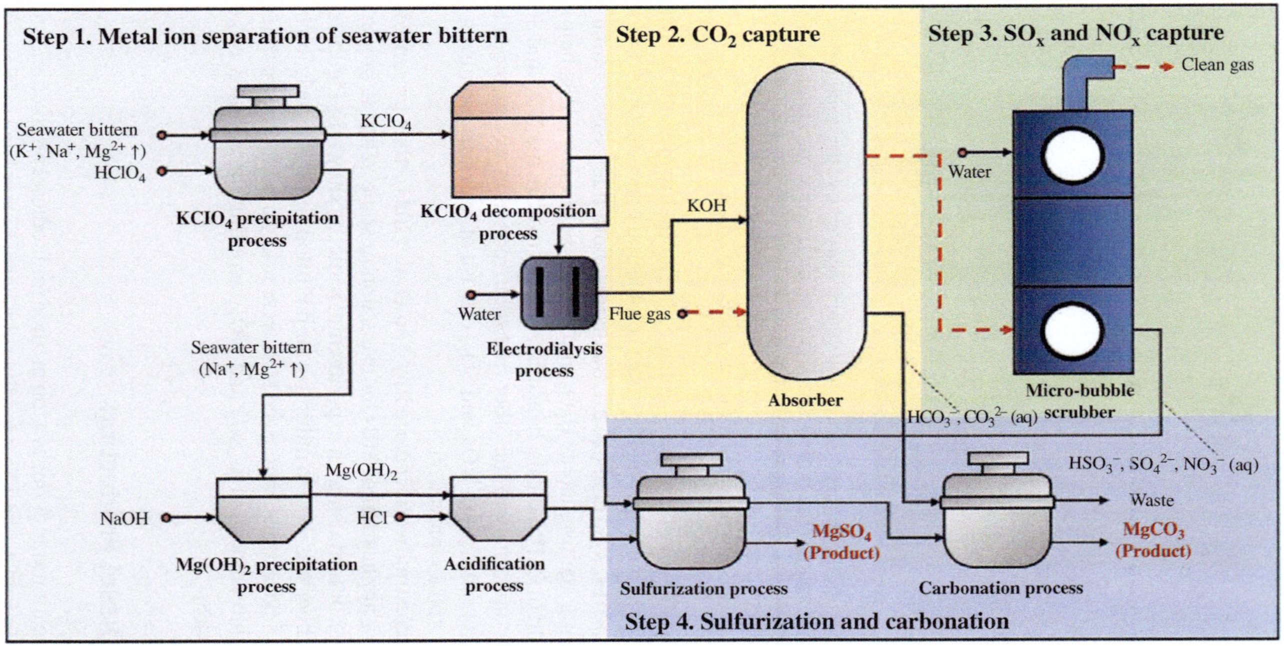

Figure 6.7 Design of experimental setup for the separation of metal ions from seawater bittern along with the desulfurization process. Source: Yoo et al. (2023)/with permission of ELSEVIER.

catalysts like H_2SO_4 and ionic liquids (ILs) due to their excellent protonation capabilities and miscibility with FFAs (Wang et al., 2020). However, this high solubility poses challenges during the purification step. While H_2SO_4 is cost-effective, it presents numerous hurdles compared to ILs, which are favored in esterification due to their low volatility, chemical stability, reactivity, and different physical and chemical properties (Ong et al., 2021). Yet, ILs face barriers like separation difficulties, high viscosity, and low diffusivity. To tackle these issues, integrating ILs as acid catalysts within porous materials such as carbon nanotubes (CNTs), metal–organic frameworks (MOFs), or zeolites is proposed. MOFs offer a significant advantage over other porous materials due to their adaptable pore structures, modifiable through organic linkers, thus enhancing the reactive species encapsulation.

MOFs show significant advantages such as a large surface area, adjustable composition, thermal and chemical stability, and straightforward preparation methods. They offer additional active sites that enhance catalyst activity, rendering them appealing as heterogeneous catalysts for biodiesel synthesis. Consequently, a catalyst based on both an IL and MOF can provide the necessary catalytic activity while facilitating separation processes for esterification. Nonetheless, acid catalysis encounters limitations due to equilibrium establishment, leading to reduced overall conversion, while the reverse reaction affects the reaction rates. In oleic acid esterification, a primary challenge arises from its miscibility in methanol (MeOH), resulting in poor reagent contact and the formation of an immiscible film that restricts mass transfer rates, thereby limiting overall reaction rates. Enhancing mass transfer, reducing film thickness, and modifying certain physical properties can improve reagent contact intimacy. Techniques such as vigorous agitation, ultrasonication, or supercritical methods can reduce film thickness, albeit at the cost of increased energy consumption, thus elevating biodiesel production expenses.

A novel MB technique has emerged to address the limitations of esterification and transesterification processes (Desai et al., 2018). This innovative approach has led to significantly improved kinetics and conversion rates of FFA to fatty acid methyl esters (FAMEs). In the MB-based method, methanol (MeOH) vapor is passed in the form of MBs into a bubble reactor containing FFA. As these MB clouds rise, they react with the FFA in the reactor. Their small size results in reduced buoyancy force that increases residence time, thereby enhancing both heat and mass transfer within the system.

6.6 Biogas Upgradation

Various techniques are applied for CO_2 removal for the production of biomethane from biogas. Adsorption including pressure swing adsorption (PSA), chemical absorption, cryogenic separation, biological methane enrichment, and membrane separation are the most preferred technologies for this. However, these technologies often require significant investments, and operating costs additionally can cause environmental pollution (Dumruangsri et al., 2021).

An interesting alternative method of gas adsorption is gas-in-water absorption, which directly depends on the solubility and contact area between the gas and liquid phases. At 30 °C, the solubilities of CO_2 and CH_4 in water are 1.25 and 0.0175 g/g H_2O, respectively, making CO_2 70 times more soluble than CH_4. Higher solubility of CO_2 is due to its different electronegativity, which, when surrounded by polar water molecules, forms a cage structure. The CO_2 removal ratio increases significantly with elevated pressure, going from 34.6% to 94.2%. Studies on water scrubbers for CO_2 and H_2S removal from biogas derived from manure showed a CO_2 removal efficiency of 21.2% and an H_2S removal efficiency of 32.8%. MBs enhance this process due to their large interfacial area of contact with the liquid, which facilitates gas adsorption. Additionally, MBs have a slow-rise velocity due to less buoyancy relative to viscous drag, resulting in a longer residence time and more effective gas absorption.

In a work reported by Paoaiang et al. (2020), a system for converting biogas to biomethane was studied in a continuous water absorption-regeneration system using MBs technology (refer to Figure 6.8). The biogas conversion system was tested with a simulated biogas mixture containing 30% CO_2 and N_2. MBs in the water as the primary absorbent were generated through a 0.5 inch venturi ejector attached before the gas separation unit. Several gas flow rates of the synthetic simulated biogas, ranging from 2 to 10 L/min, were tested for operation, having a constant water flow rate of 15 L/min and operating pressure of 7 bar while monitoring CO_2 %removal efficiency.

The regeneration unit was a counter-current flow unit between air and water, with airflow rates ranging from 5 to 30 L/min. The optimum absorption conditions were at a liquid–gas ratio of 7.5, which was achieved with a biogas flow rate of 2 L/min and a water flow rate of 15 L/min. An optimum air flow rate of 30 L/min in the regeneration unit resulted in a CO_2 %removal efficiency of over 90% from the simulated biomethane.

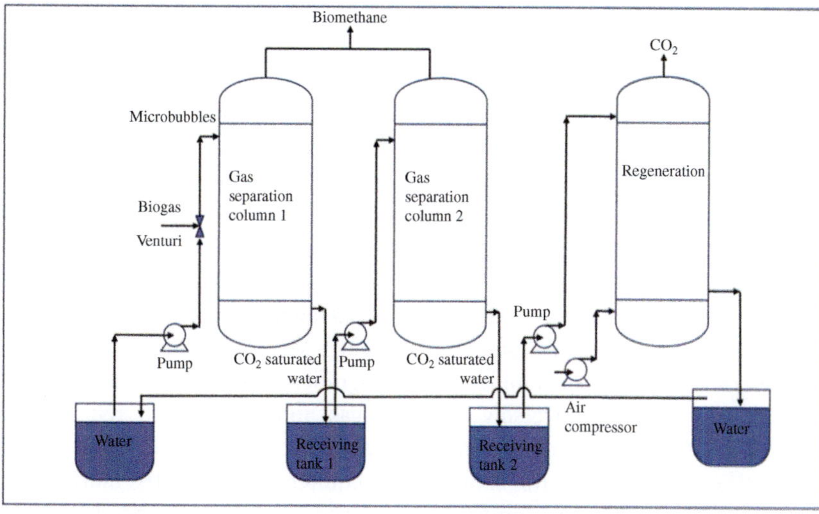

Figure 6.8 Schematic flow diagram of CO_2 absorption and gas separation using MBs. Source: Adapted from Paoaiang et al. (2020).

6.7 Enhanced Oil Recovery by CO_2-Microbubbles (MBs)

In CO_2-enhanced oil recovery (EOR), the injection of CO_2 into oil reservoirs can be taken place by two different methods, either through miscible CO_2 with oil or immiscible CO_2 and oil method, both aiming to augment oil extraction rates. This conventional approach also involves utilizing geological formations as sinks for CO_2 sequestration. CO_2 gas has shown significant potential for both EOR and sequestration (Xue et al., 2014; Telmadarreiea et al., 2016; Su et al., 2013; Zhao and Liao, 2012; Liu et al., 2021). However, the CO_2-EOR process has drawn several drawbacks, including insufficient sweep efficiencies, suboptimal CO_2 utilization factors, and unfavorable mobility ratios. The high mobility of gas, combined with reservoir heterogeneity and gravity segregation, often leads to premature breakthrough of gas and inefficient sweep. Moreover, irregular front interfaces, caused by variations in vertical permeability, result in uneven displacement of crude oil, exacerbated by differences in horizontal permeability.

CO_2 foam presents a promising solution to mitigate the challenge of high gas mobility within porous media. The injected foam exhibits increased viscosity, resembling a more viscous fluid, which enhances displacement efficiency compared to pure gas injection (Rossen, 1996). However, the application of foam is constrained in highly fractured or heavy oil reservoirs, where high reservoir heterogeneity and elevated crude oil viscosity diminish the efficacy of foam-based EOR techniques. While CO_2 flooding is typically employed in deeper, hotter, and high-pressure reservoirs, shallow heavy oil reservoirs offer the potential for the application of CO_2 MBs, given their lower temperature and pressure profiles relative to deeper reservoirs. In deep, high-temperature, and high-pressure reservoirs, the generation of MBs necessitates careful selection of surfactants and polymers tailored for such conditions (refer Figures 6.9 and 6.10). Moreover, the adverse impact of crude oil on foam stability poses a significant challenge in foam-based EOR processes. Conversely, colloidal gas aphrons (CGAs) represent a stable MBs system incorporating surfactants, polymers, and gas within their structure. Since the components required to create MBs (surfactant, polymer, and gas) are commonly employed in EOR field operations, the integration of suitable polymers and surfactants with CO_2 gas enhances the stability of MBS and the bubble mobility-reducing properties. Consequently, a comprehensive understanding of the stability and flow characteristics of MBs is imperative for the successful deployment of CO_2 MBs in EOR and sequestration applications.

6.8 Microbubbles (MBs)-Membrane Systems

The combination of membrane process with the MBs could be used for three different purposes, such as

(i) Generation of MBs and NBs using the porous membranes as a replacement for gas spargers

(ii) Use of MBs for membrane cleaning and de-fouling

(iii) Simultaneous use of MBs and membranes for enhancing the efficiency of the process in any stage of water treatment; however, mostly the tertiary treatment process

6.8.1 Membrane for Microbubble Generation

Kukizaki and his coworkers (Kukizaki and Goto, 2007; Kukizaki and Baba, 2008; Kukizaki, 2009) successfully developed a glass membrane capable of generating uniform-sized MBs. The image of their ceramic membrane is shown in Figure 6.11. They observed a significant reduction in the average MB diameter with increasing liquid flow rate and decreasing liquid

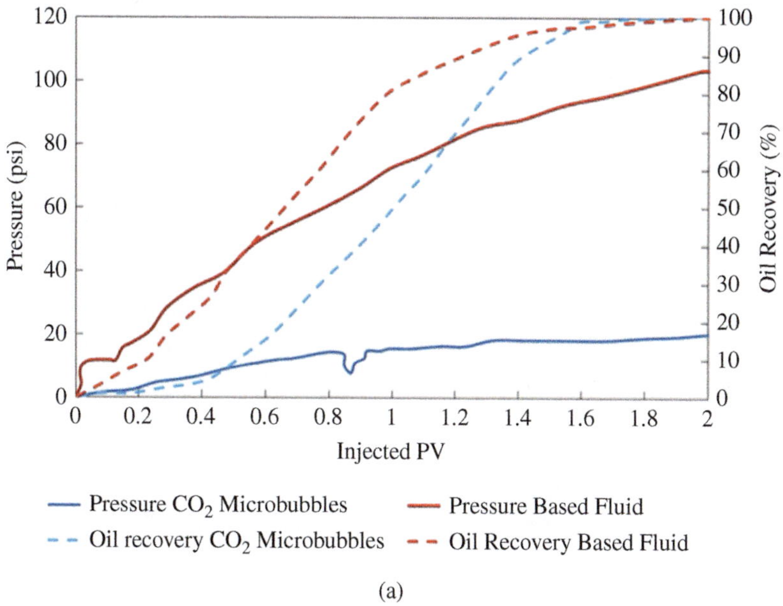

Figure 6.9 (a) Effect of injection pressure for oil recovery using CO_2 MBs and (b) the photograph of sample containing CO_2 MBs. Source: Reproduced with permission from Telmadarreiea et al. (2016)/ELSEVIER.

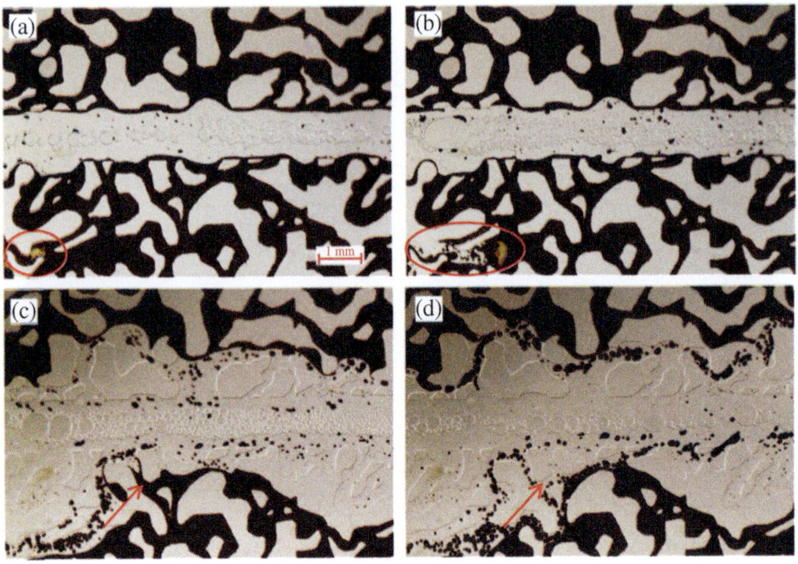

Figure 6.10 The images of heavy oil recovery using CO_2 MBs, taken (a) after 15 min, (b) after 20 min, (c) after 60 min, and (d) after 65 min. The early stages after 15–20 min represent the oil suspension, whereas later stages show flooding. Source: Reproduced with permission from Telmadarreie et al., 2016/ELSEVIER.

surface tension. Numerous studies in the literature have explored the controllable generation of MBs, investigating the impact of physical properties such as viscosity, surface tension, liquid and gas flow rate, and membrane pore size on the size of MBs. For instance, liquid viscosity exerts a dual effect on MB size when using a multichannel ceramic membrane (Liu et al., 2018). Initially, as liquid viscosity increased, the d_{32} of MBs decreased at $\mu < 2$ mPa·s, followed by a gradual increase of μ from 2.0–20 mPa·s. Xie et al. (2021a) also verified that the size of MBs generated in aqueous solutions through a ceramic membrane was significantly influenced by flow rate of liquid, viscosity, and surface tension. Additionally, the average size of MBs was found to be affected by the pores size of the membrane and, more importantly, the properties of surfactants.

Zhu et al. (2010) studied the growth and detachment of MBs at the tips of capillary tubes using a high-speed camera. Interestingly, they observed that the size of MBs formed at the tip of a hydrophobic capillary (of PTFE material) at a contact angle of 93° was significantly lower than those formed at a hydrophilic glass capillary (with a contact angle of 37°). Several research groups have agreed that the formation of bubbles on an orifice plate is indeed influenced by surface wettability.

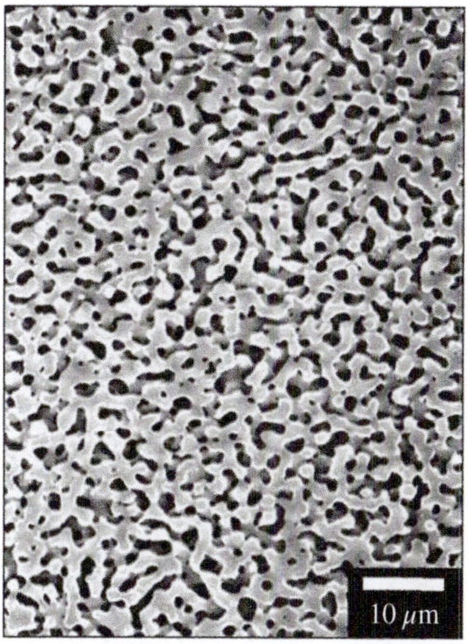

Figure 6.11 The morphology of SPG membrane contactor for generating MBs. Source: Reproduced with permission from Kukizaki et al. (2007)/ELSEVIER.

6.8.2 Modeling of Microbubble Size Generated from Porous Membrane

Xie et al. (2021a, 2021b) have investigated several operating parameters influencing the size of MBs, including liquid and gas flow rate, surface tension, and viscosity. It was observed that in aqueous solutions, MB size decreased with increasing liquid flow rate and decreasing liquid surface tension, consistent with findings in the literature. The liquid viscosity exhibited a dual effect on MB size while using a hydrophilic membrane. To predict MB size, a monodispersion model was developed, as the MBs tend to originate from membrane pores in the presence of surfactants. On the contrary, in organic solvents, MBs were generated from the hydrophilic membrane, from several pores adjacent to each other due to MBs expansion at the three-phase system, rendering the monodispersion model inadequate for predicting MBs formation in such solvents.

Xu et al. (2006, 2008) introduced a mathematical model to predict droplet size in T-junction microchannels of width w_m and height h_m, operating under a dripping regime. Their findings highlighted the dominance of the balance among the shear stress and interfacial force in droplet breakup, which was extended to elucidate MB formation through porous membranes, having the length of dispersed droplet l.

$$\frac{l}{w_m} - \mathrm{PI} = k\left(\frac{Q_d}{Q_c}\right)^{\alpha}\left(\frac{1}{C_a}\right)^{\beta} \tag{6.21}$$

The PI is calculated as follows:

$$\mathrm{PI} = \frac{1}{D_{av}}\sqrt{\frac{1}{n-1}\sum_{i=1}^{n}(D_i - D_{av})^2} \times 100\% \tag{6.22}$$

where D_i is the size of the ith MB of the total number of bubbles n and D_{av} is the average MB diameter.

Mostly, the equivalent pore diameter is used to represent the effect of several adjacent pores on MB size as follows:

$$\frac{D_{av}}{d_i} = k\left(\frac{Q_g}{Q_l}\right)^{\alpha}\left(\frac{1}{C_a}\right)^{\beta} \tag{6.23}$$

$$\frac{D_{av}}{d_i} = 1735.67\left(\frac{Q_g}{Q_l}\right)^{0.74}\left(\frac{1}{C_a}\right)^{-0.17} \tag{6.24}$$

The values of constants k, α, and β were calculated and represented in Eq. (6.24), C_a is the capillary number for continuous phase, which is also the ratio of viscous and interfacial tension stress, and Q_g and Q_l are the gas and liquid flow rates, respectively. The effect of pore size of membrane on the size of the MBs generated has been compared by Xie et al. (2021b) and presented in Figure 6.12. The experimental and predicted values from the model for the D_{av} were also compared and found to have a good match, as shown in Figure 6.13. Membrane wettability plays a pivotal role in determining MB size. Notably, MBs generated in organic solvents using hydrophilic ceramic membranes exhibit larger sizes compared to those formed in water. This discrepancy arises from the tendency of MBs in organic solvents to expand

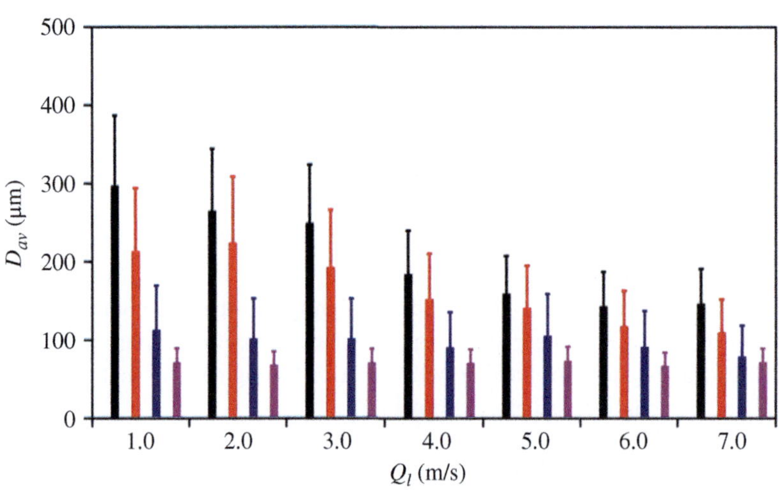

Figure 6.12 Membrane pores controlling the MB diameter in the ceramic membrane contactors. Source: Xie et al. (2021b)/American Chemical Society.

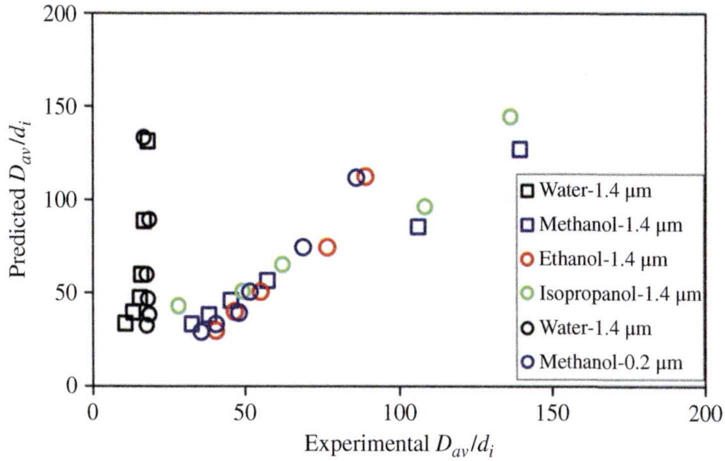

Figure 6.13 Matching the experimental and predicted data for the MBs average diameter obtained in the ceramic membrane contactor. Source: Adapted from Xie et al. (2021b).

at the three-phase line, originating from larger equivalent pores. Conversely, MBs generated in hydrophobic PTFE membranes exhibit an inverse trend. Additionally, the size of MBs is markedly influenced by liquid flow rate and viscosity.

6.8.3 Membrane Cleaning

The utilization of MBs for membrane cleaning has garnered considerable attention in recent years (Ghadimkhani et al., 2016; Watabe et al., 2016, 2018; Harun and Zimmerman, 2019; Wang et al., 2021; Hashimoto et al., 2022). Many researchers have explored the efficacy of MB in membrane washing. For instance, Ghadimkhani et al. (2016) employed air ultrafine bubbles for cleaning ceramic membranes fouled with humic acid, observing a restoration of the membranes' initial flux after an extended cleaning duration of 120 minutes. Their method involved generating air ultrafine bubbles through a hybrid approach combining pressurized-dissolution and swirling-liquid flow methods, resulting in simultaneous generation of MBs alongside air ultrafine bubbles, contributing to the observed cleaning effect. Similarly, Wang et al. (2021) utilized MB to remove oil foulants following the separation of an oil-in-water emulsion with polyacrylonitrile carbide membranes doped with manganese dioxide. In their cleaning protocol, H_2O_2 was separated through the membrane with oxygen MBs, while manganese dioxide acted as the catalyst to enhance cleaning efficacy. Although the authors reported a 67% recovery of initial flux, it is noted that this result might include the cleaning effect of hydrogen peroxide itself, with any specific impact of MBs on fouling suppression not clearly discernible.

Fine bubbles (FBs) are believed to clean fouling through one or a combination of the several mechanisms as follows:

1) **Cake layer detachment:** Facilitated by the shearing action of FBs (Watabe et al., 2016).
2) **Adsorption and surfacing:** Foulants adhere to the FB surface and are carried away.
3) **Dislodgment:** Generated by FB disintegration, which also leads to organic matter decomposition by hydroxyl radicals.
4) **Augmentation of foulant adsorption:** The interaction between the zeta potential of foulants and FBs, along with the modification of foulant membrane surface interaction, alters the foulant's zeta potential.

Conventional scrubbing primarily removes foulants through the shear force exerted by bubbles impacting the membrane surface. This mechanism takes into account bubble adsorption onto foulants and their removal by bubbles, though the extent of this contribution is unclear. Thus, applying FB cleaning to partially scrubbed membranes enhances the scrubbing effect. In the case of monolithic ceramic membranes that are unsuitable for traditional scrubbing, FB cleaning introduces a novel removal mechanism to improve overall cleaning efficiency.

6.8.4 Simultaneous Use of Microbubbles and Membranes for Water Treatment

Reverse osmosis is extensively utilized in seawater desalination and industrial water reclamation due to its robust capacity to remove a range of contaminants, salts, and pathogens. However, the byproduct of reverse osmosis, known as reverse osmosis concentrate (ROC), contains complex recalcitrant pollutants and salts, posing significant challenges for subsequent treatment and discharge. Catalytic ozonation, which relies on hydroxyl radical generation, has gained widespread attention for its effectiveness in eliminating toxic organics from wastewater, offering promising prospects for ROC treatment. Despite transitional metal oxides being common catalysts in ozonation processes, issues such as metal leaching, nanoparticle agglomeration, and their poor recovery significantly hinder their practical application on a large scale.

Several studies have explored combining O_3-MBs with Al_2O_3 ceramic membranes, indicating that MBs can effectively reduce fouling adhesion to the membrane. Specifically, ozone MBs have shown the ability to penetrate pollution layers and generate hydroxyl radicals on Al_2O_3 ceramic membranes, further disrupting the pollution layer (Duan et al., 2022; Han et al., 2017).

Integrating membrane filtration with MBs offers a promising approach for treatment of dye wastewater due to its environmental compatibility and efficacy. However, the generation of hydroxyl radicals during MBs self-dissolution may be insufficient for practical applications. This study introduces the characterization of a modified Al_2O_3 ceramic membrane (CM) incorporating MnO_2 nanosheets (MnO_2/Al_2O_3-CM), providing dual functions of membrane filtration and MBs catalysis. A novel MB catalytic membrane reactor (MNB-CMR) is constructed using the modified MnO_2/Al_2O_3-CM to treat simulated dye wastewater, specifically methylene blue. The dye degradation profile shows that the MB-CMR improves decolorization rates by 17% and TOC removal rates by 20% compared to the integrated MBs process with the non-modified CM. The MB-CMR demonstrates excellent catalytic activity and fouling resistance. Additionally, the study investigates the effects of methylene blue concentration, temperature, solution flux, and pH on dye degradation performance.

Some more applications of MBs have been listed in Table 6.1 to get a comprehensive knowledge of the versatile technology in some nontraditional processes.

Table 6.1 Some nontraditional applications of MBs to enhance the process efficiency.

Process/Industry	Objective	MB gas phase	References
Reactive crystallization	$CaCO_3$ crystallization	CO_2/NH_3	Matsumoto et al. (2010)
Esterification	Ethyl acetate production	Air	Javed et al. (2019).
Beauty and cosmetic	Human skin cleaning	Air	Kang et al. (2024)
Decarbonization of air	CO2 reaction with amines	CO_2	Paiva (2022)
Fish farming	Promoting fish health	O_2	Heriyati et al. (2022)
Foam production	Natural rubber latex foams	Air, O_2	Katkeaw et al. 2022
Activated sludge process	Enhance aerobic digestion of activated sludge processing	Air, O_2	Zhou et al. (2022)
Food processing	Cleaning and disinfection	Air, O_2, N_2	Lu et al. (2023)
Algal bioprocessing	Microalgal growth	Air, O_2	Patel et al. (2021)
Sediment and land remediation	Improving soil quality, free from heavy metals	Air, O_2	Sha et al. (2020)

Nomenclature

Notation		SI Unit
C_a	capillary number for continuous phase	unit less
C_F	concentration of target minerals in feed	mol/L
C_{LO}	concentration of target minerals in scrubbing liquor	mol/L
C_M	concentration of target minerals in supernatant	mol/L
C_R	concentration of target minerals in raffinate	mol/L
C_{SC}	concentration of target minerals in scrubbing liquor	mol/L
C_{ST}	concentration of target minerals in stripping liquor	mol/L
d_{10}	10% of the cumulative particle size in the size distribution curve	m
d_{50}	50% of the cumulative particle size in the size distribution curve	m
d_{90}	90% of the cumulative particle size in the size distribution curve	m
d_b	bubble diameter	m
d_p	particle size	m
d_F	fine particle size	m
D_{c1}	distribution coefficient	unit less
D_{Coeff}	distribution coefficient	unit less
D_{MB}	MBs density in the column	number/m^3
D_{av}	average MB diameter	m
D_c	column diameter	m
E_a	adhesion efficiency	%
E_c	particle collision efficiency	%
E_s	stability efficiency	%
F_B	buoyancy force	N
F_D	drag force	N
F_G	gravitational force	N
g	gravitational constant	m/s^2
h	height under dripping regime	m

Notation		SI Unit
h_m	T-junction microchannels height	m
k_1	constant	unit less
K_F	rate constant for the particle–MBs flotation	concentration$^{1\text{-order}}$/time
l	length of dispersed droplet	m
L	flow length scale	m
m	mass of particle or bubble	kg
N_p	number density of fine floatable particle	unit less
PI	polydispersity index	unit less
Q_c	flow rate of continuous phase	m^3/s
Q_d	flow rate of dispersed phase	m^3/s
Q_g	gas flowrate	m^3/s
Q_l	liquid flowrate	m^3/s
R_c	recovery of fine particles	%
R_{max}	maximum possible recovery	%
t	time	s
u_p	particle rise velocity	m/s
$U_{c,x}$	particle–MBs mixture circulation velocity in x axis	m/s
$U_{c,y}$	particle–MBs mixture circulation velocity in y axis	m/s
$U_{c,z}$	particle–MBs mixture circulation velocity in z axis	m/s
V_F	volume of feed	m^3
V_{LO}	volume of target minerals in floating organic phase	m^3
ν_p	dynamic viscosity of particle	Pa.s
V_R	volume of raffinate	m^3
V_{SC}	volume of target minerals in scrubbing liquor	m^3
V_{ST}	volume of target minerals in stripping liquor	m^3
ν_L	fluid kinematic viscosity	m^2/s
ν	velocity of particle or bubble	m/s
V_F	root mean square velocity of fine particles	m/s

Notation		SI Unit
V_{MBs}	root mean square velocity of MBs	m/s
w	weight under dripping regime	kg
w_m	T-junction microchannels width	m
Z	collision frequency	m^3/s

Greek Nomenclature

Notation	SI Unit
α	constant
β	constant
v_p	dynamic viscosities of the particle
ε	rate of energy dissipation per unit mass
ϵ_G	gas volume fraction
η_M	mineralization efficiency
η_{SC}	scrubbing efficiency
η_{ST}	stripping efficiency
μ_p	kinematic viscosity of particle
μ_l	kinematic viscosity of fluid
ρ_F	density of fluid
ρ_L	liquid density
ρ_p	particle density
$\Delta\rho$	difference in density of fine particle

References

Ahmed, N. and Jameson, G.J. (1985). The effect of bubble size on the rate of flotation of fine. Int. J. Miner. Process. 14: 195–215.

Bloom, F. and Heindel, T.J. (1997). Mathematical modeling of the flotation deinking process. Math. Comput. Model. 25: 13–58.

Desai, P.D., Hines, M.J., Riaz, Y., and Zimmerman, W.B. (2018). Resonant pulsing frequency effect for much smaller bubble formation with fluidic oscillation. Energies 11: 2680.

Dobby, G.S. and Finch, J.A. (1987). Particle size dependence in flotation derived from a fundamental model of the capture process. Int. J. Miner. Process. 21: 241–260.

Duan, Y., Yu, J., Zhang, R. et al. (2022). Integrated MnO_2 nanosheet ultrafiltration ceramic membrane with micro-nano bubbles for catalytic treatment of dye wastewater. Sep. Purif. Technol. 300: 121786.

Dumruangsri, C., Pongyeela, P., and Juntima, C.J. (2021). Biogas upgrading to biomethane by CO_2 removal using water absorber with microbubble technique. Walailak J. Sci. Tech. 18: 9304.

Etchepare, R., Oliveira, H., Nicknig, M. et al. (2017). Nanobubbles: generation using a multiphase pump, properties and features in flotation. Min. Eng. 112: 19–26.

Gontijo, D.F.C., Fornasiero, D., and Ralston, J. (2007). The limits of fine and coarse particle flotation. CJCE 2007 (85): 739–747.

Ghadimkhani, A., Zhang, W., and Marhaba, T. (2016). Ceramic membrane defouling (cleaning) by air Nano bubbles. Chemosphere 146: 379–384.

Han, Y., Liu, Y., Jiang, H. et al. (2017). Large scale preparation of microbubbles by multi-channel ceramic membranes: hydrodynamics and mass transfer characteristics. CJCE 95: 2176–2185.

Harun, M.H.C. and Zimmerman, W.B. (2019). Membrane defouling using microbubbles generated by fluidic oscillation. Water Supply 19: 97–106.

Hashimoto, K., Onzuka, A., Nishijima, W. et al. (2022). Effect of fine bubbles for washing of monolith type porous ceramic membranes treating oil-in-water emulsions. Chemosphere 305: 135487.

Heriyati, E., Rustadi, R., Isnansetyo, A. et al. (2022). Microbubble aeration in a recirculating aquaculture system (RAS) increased dissolved oxygen, fish culture performance, and stress resistance of red tilapia (Oreochromis sp.). Trends Sci. 19: 6251.

Javed, F., Shamair, Z., Ali, S. et al. (2019). "Pushing and pulling" the equilibrium through bubble mediated reactive separation for ethyl acetate production. React. Chem. Eng. 4: 705–714.

Jung, M.U., Kim, C.Y., Bournival, G., and Ata, S. (2023). Industrial application of microbubble generation methods for recovering fine particles through froth flotation: a review of the state-of-the-art and perspectives. Adv. Collod. Inter. Sci. 322: 103047.

Kang, Y., Park, J., and Park, H. (2024). Toward the eco-friendly cosmetic cleansing assisted by the micro-bubbly jet. Sci. Rep. 14: 8189.

Katkeaw, K., Khangkhamano, M., and Kokoo, R. (2022). Microbubble technology for natural rubber latex foam production: the use of various gas-filled microbubbles. Cell. Polym. 41: 21–29.

Kukizaki, M. (2009). Microbubble formation using asymmetric Shirasu porous glass (SPG) membranes and porous ceramic membranes-a comparative study. Colloids Surf. A 340: 20–32.

Kukizaki, M. and Baba, Y. (2008). Effect of surfactant type on microbubble formation behavior using Shirasu porous glass (SPG) membranes. Colloids Surf. A 326: 129–137.

Kukizaki, M. and Goto, M. (2007). Spontaneous formation behavior of uniform-sized microbubbles from Shirasu porous glass (SPG) membranes in the absence of water-phase flow. Colloids Surf. A 296: 174–181.

Li, P., Wang, H.J., Ueda, Y. et al. (2022). Microbubbles for effective cleaning of metal surfaces without chemical agents. Langmuir 38: 769–776.

Liu, Y., Han, Y., Li, X. et al. (2018). Efficient control of microbubble properties by alcohol shear flows in ceramic membrane channels. Chem. Eng. Technol. 41: 168–174.

Liu, Y., Tao, X., Jiang, H., and Chen, R. (2020). Intensification of fine apatite flotation with microbubble generation and inclined plates in the flotation column. Chem. Eng. Process. Proc. Inten. 157: 108133.

Liu, N., Chen, X., Ju, B. et al. (2021). Microbubbles generation by an orifice spraying method in a water-gas dispersion flooding system for enhanced oil recovery. J. Petrol. Sci. and Eng. 198: 108196.

Lu, J., Owen, G., Jones, G.O. et al. (2023). Microbubbles in food technology, annual revi. of Food Sci. and Technol. 14: 495–515.

Matsumoto, M., Fukunaga, T., and Onoe, K. (2010). Polymorph control of calcium carbonate by reactive crystallization using microbubble technique. Chem. Engine. Res. Des. 88: 1624–1630.

Miyamoto, M., Ueyama, S., Hinomoto, N. et al. (2007). Degreasing of solid surfaces by microbubble cleaning. Jap. J. Appl. Phys. 46: 1236–1243.

Oliveira, H., Azevedo, A., and Rubio, J. (2018). Nanobubbles generation in a high-rate hydrodynamic cavitation tube. Min. Eng. 116: 32–34.

Ong, C.H., Tiong, W.Y., Goh, H.H.B. et al. (2021). Recent advances in biodiesel production from agricultural products and microalgae using ionic liquids: opportunities and challenges. Energy Conv. And Manag. 228: 113647.

Paiva, S.L. (2022). Boosting reactions with microbubbles. Nat. Rev. Chem. 6: https://doi.org/10.1038/s41570-021-00355-5.

Paoaiang, J., Dumruangsri, C., Chaisongkroh, N. et al. (2020). Continuous water absorption - regeneration using microbubble technology for CO_2 removal in biogas upgrading system. Inter. Energy J. 20: 629–636.

Parmar, R. and Majumder, S.K. (2016). Mineral beneficiation by ionic microbubble in continuous plant prototype: efficiency and its analysis by kinetic model. Chem. Eng. Sci. 142: 42–54.

Patel, K.A., Singhania, R.R., Chen, W.C. et al. (2021). Advances in micro- and nano bubbles technology for application in biochemical processes. Envir. Technol. Inov. 23: 101729.

Pope, S.B. (2000). *Turbulent Flows*. Cambridge, UK: Cambridge University Press.

Pyke, B., Fornasiero, D., and Ralston, J. (2003). Bubble particle hetero coagulation under turbulent conditions. J. Colloid Interface Sci. 2003 (265): 141–151.

Ralston, J., Fornasiero, D., and Hayes, R. (1999). Bubble–particle attachment and detachment in flotation. Int. J. Miner. Process. 56: 133–164.

Rosa, F.A. and Rubio, J. (2018). On the role of nanobubbles in particle–bubble adhesion for the flotation of quartz and apatitic minerals. Mineral. Eng. 127: 178–184.

Rossen, W.R., 1996. Foams in enhanced oil recovery, in foams: theory. In: Prud'- homme, R.K., Khan, S.A. (Eds.), measurements and applications. Marcel Dekker Inc., New York. 1996.

Ruoyi, Y., Fujii, K., Kaneko, A. et al. (2020). Degreasing technology with ozone micro-bubbles generated by a venturi tube. Multi. Sci. and Technol. 32: 47–60.

Sha, Z., Chen, Z., Feng, Y. et al. (2020). Minerals loaded with oxygen nanobubbles mitigate arsenic translocation from paddy soils to rice. J. Hazard. Mater. 398: 122818.

Su, K., Lia, X., Zhao, X., and Zhang, H. (2013). Coupled CO_2 enhanced oil recovery and sequestration in China's demonstration project: case study and parameter optimization. Energy Fuels 27: 378–386.

Su, S., Huang, Y., Liu, B. et al. (2022). A feasible strategy for deeply separating low concentrations of molybdenum from tungstate solutions with a high-efficiency microbubble floating-extraction concept. ACS Sustain. Chem. Eng. 10: 146–158.

Tan, A.K., Mohan, Y., Liew, J.K. et al. (2020). Development of an effective cleaning method for metallic parts using microbubbles. J. Clean. Prod. 261: 121076.

Telmadarreiea, A., Dodaa, A., Trivedia, J.J. et al. (2016). CO_2 microbubbles– a potential fluid for enhanced oil recovery: bulk and porous media studies. J. Pet. Eng. 138: 160–173.

Tsave, P.K., Kostoglou, M., Karapantsios, T.D., and Lazaridis, N.K. (2021). A hybrid device for enhancing flotation of fine particles by combining micro-bubbles with conventional bubbles. Minerals 11: 561.

Wang, S., Yan, W., and Zhao, F. (2020). Recovery of solid waste as functional heterogeneous catalysts for organic pollutant removal and biodiesel production. Chem. Eng. J. 401: 126104.

Wang, Y., Wang, J., Ding, Y. et al. (2021). In situ generated micro-bubbles enhanced membrane antifouling for separation of oil-in-water emulsion. J. Membr. Sci. 621: 119005.

Watabe, T., Matsuyama, K., Takahashi, T., and Matsuyama, H. (2016). The effect of microbubbles on membrane fouling caused by different foulants. Desalin Water Treat. 57: 9558–9568.

Watabe, T., Takahashi, T., Matsuyama, K., and Matsuyama, H. (2018). Effect of microbubbles on membrane fouling due to protein in water treatment processes. Desalin Water Treat. 2018 (120): 9–15.

Weber, M.E. and Paddock, D. (1983). Interceptional and gravitational collision efficiencies for single collectors at intermediate Reynolds numbers. J. Colloid Interface Sci. 94: 328–335.

Wu, H. and Patterson, G.K. (1989). Laser-doppler measurements of turbulent-flow parameters in a stirred mixer. Chem. Eng. Sci. 44: 2207–2221.

Xiao, Z. and Li, D. (2019). Simultaneous removal of NO and SO_2 with a micro-bubble gas-liquid dispersion system based on air/H_2O_2/$Na_2S_2O_8$. Env. Technol. 41: 3573–3583.

Xiao, Z., Li, D., Zhu, Q., and Sun, Z. (2020). Simultaneous removal of NO and SO_2 through a new wet recycling oxidation–reduction process utilizing micro-nano bubble gas–liquid dispersion system based on Na_2SO_3. Fuel 263: 116682.

Xie, B.Q., Zhou, C.J., Sang, L. et al. (2021a). Preparation and characterization of microbubbles with a porous ceramic membrane. Chem. Eng. Process. 159: 108213.

Xie, B., Zhou, C., Huang, X. et al. (2021b). Microbubble generation in organic solvents by porous membranes with different membrane wettabilities. Ind. Eng. Chem. Res. 60: 8579–8587.

Xu, J.H., Li, S.W., Chen, G.G., and Luo, G.S. (2006). Formation of monodisperse microbubbles in a microfluidic device. AIChE J. 52: 2254–2259.

Xu, J.H., Li, S.W., Tan, J., and Luo, G.S. (2008). Correlations of droplet formation in T-junction microfluidic devices: from squeezing to dripping. Microfluid Nanofluid 5: 711–717.

Xue, Z., Nishio, S., Naoto, H.N., and Matsuoka, T. (2014). Microbubble carbon dioxide injection for enhanced dissolution in geological sequestration and improved oil recovery. Energy Procedia 63: 7939–7946.

Yoo, Y., Lim, J., Kim, J., and Cho, H. (2023). Seawater bittern recovery system for CO_2, SOx and NOx removal using microbubble scrubber. Desalination 558: 116612.

Zhao, X. and Liao, X. (2012). Evaluation method of CO_2 sequestration and enhanced oil recovery in an oil reservoir, as applied to the change in oilfields. China. Energy Fuels 26: 5350–5354.

Zhu, X., Liao, Q., Wang, H. et al. (2010). Experimental study of bubble growth and departure at the tip of capillary tubes with various wettabilities in a stagnant liquid. J. Supercond. Novel Magn. 23: 1141–1145.

7

CFD and Microbubbles

Dr Abhishek Yadav

Assistant Professor, Department of Chemical Engineering, Pandit Deendayal Energy University, Gandhinagar, Gujarat, India

7.1 Introduction

A computational fluid dynamics (CFD) study involves numerically solving a system of partial differential equations (PDEs) on a domain with appropriate boundary conditions. The CFD-assisted solutions are representative of the physical phenomena if the governing PDEs capture the physics properly. CFD analysis can help in designing an optimized experimental setup to study a multiphase flow problem by predicting the extent of various modes of transport such as mass, momentum, and heat. One can run a CFD test on a hypothetical design of a setup to find whether expected outcomes are achieved before physically fabricating the setup, thus saving both time and expenses. A parametric study in CFD analysis can also help in choosing suitable design parameters. This chapter deals with the various mathematical models (or PDEs) that govern the physics of multiphase flows.

7.2 CFD Simulation

Fluid flows are modeled by considering various local instant variables such as velocity field, v(x, t), temperature field, T(x, t), and concentration field, c(x, t). These variables originate from a continuum hypothesis that allows one to treat the fluid as being uniformly spread over a volume element rather than being concentrated on discrete and randomly changing locations in the volume element. Fluids are made of atoms or molecules that exhibit an ever-occurring random motion for any temperature above absolute zero. Therefore, any quantity, such as mass, should be located where

Microbubbles: Engineering Aspects and Industrial Applications, First Edition. Snigdha Khuntia.

the molecules are in the volume element, thus leaving a complete vacuum at the rest of the locations of the volume element. Therefore, it is impossible to define quantities such as mass and velocity at every location and time. The continuum hypothesis makes use of a huge difference in molecular motion scales (such as length and time scales) and the system scales and proposes that the fluids can be treated as a continuum with respect to the system scales. The Knudsen number is defined as $= \frac{\lambda}{\ell}$, where λ is the average mean free path of molecules and ℓ is a characteristic length scale of the system over which macroscale variables are changing. The continuum hypothesis is valid for $Kn \ll 1$ condition. Pope (2000) provides a very neat description of setting up local instant variables, such as density field and velocity field, by realizing that for $Kn \ll 1$, there exists an intermediate length scale ℓ^* such that it is very large in comparison with molecular length scales but still smaller than the system length scale, i.e. $\lambda < \ell^* < \ell$. The continuum variables or the local instant variables are nothing but molecular properties averaged over a volume of size ℓ^{*3}. This brief discussion on continuum hypothesis proves useful when hypothesizing the discrete dispersed phase particles as a continuum phase in the case of a two-fluid flow model.

The equations governing the variations of local instant variables are found by conservation principles of mass, momentum, and energy; follow Bird et al. (2002) for more details, and note that the sign convention for the shear stress tensor adopted here is different from this reference. For a single-phase flow, the mass conservation results in the continuity equation (Eq. (7.1)), the momentum conservation results in the Navier–Stokes equations (Eq. (7.2)), and the first law of thermodynamics results in the equation governing the energy/temperature field (Eq. (7.3)) of the fluid. These equations have been referred to as local instant equations throughout this text.

$$\frac{\partial \rho}{\partial t} + \nabla \cdot (\rho \mathbf{v}) = 0 \tag{7.1}$$

$$\begin{aligned} &\rho\left(\frac{\partial v}{\partial t} + v \cdot \nabla v\right) = -\nabla p + \nabla \cdot \tau + \rho g, \\ &\tau = 2\mu S - \frac{2}{3}\mu(\nabla \cdot v), \\ &S = \frac{1}{2}(\nabla v + \nabla v^T) \end{aligned} \tag{7.2}$$

$$\begin{aligned} &\rho\left(\frac{\partial H}{\partial t} + v \cdot \nabla H\right) = -\nabla \cdot q + \tau : \nabla v + \left(\frac{\partial p}{\partial t} + v \cdot \nabla p\right), \\ &q = -\lambda \nabla T, \lambda \equiv \textit{Thermal conductivity} \end{aligned} \tag{7.3}$$

These equations are usually coupled PDEs and should be solved simultaneously to find out the solution in terms of velocity, pressure, temperature, and concentration fields. Finding close-form solutions is difficult even for relatively simpler flow domain geometries; therefore, usually, these equations are solved numerically with the help of digital computers.

7.2.1 Two-Phase Flow Models

Multiphase flows are characterized by the presence of at least two or more phases containing single or multiple chemical species in the flow domain. For example, the flow of gas bubbles in a pool of liquid; here the liquid phase encompasses many distinct gas bubbles inside it; therefore, the liquid is called the continuous phase and the gas is called the dispersed phase. In our discussion, we consider only two-phase flows, but all the equations presented henceforth are equally valid for more than two-phase cases. Another terminology adopted in this discussion is that the disconnected regions of dispersed phase are called particles or bubbles.

The first approach to model multiphase flows is to apply the continuum-based local instant models described in Eqs. (7.1)–(7.3) separately in each particle domain as well as in continuous phase domain. The moving interfaces of particles are modeled using so-called jump conditions (Delhaye, 1974). For a detailed discussion on the jump condition. Such an approach is very fundamental but requires solving a very large number of equations. In a typical bubble-in-water flow case, there are approximately 10^6 microbubbles in 1 mL of water. Solving such a large number of equations and resolving all the length and time scales make the problem intractable. Therefore, different kinds of approximations have been employed to construct mathematical models that can provide reasonably good solutions and yet require relatively less computational time and resources.

7.2.1.1 Two-Fluid Model

The local instant equations for a single-phase flow case are developed by realizing the continuum of the fluid on a length and time scale larger than the molecular motion scales. This idea results in local instant variables that are modeled using various conservation laws as described in the previous section. An extrapolation of the continuum hypothesis could be visualizing the collection of dispersed phase particles moving in a "random" motion as another continuum but with a larger length and time scale. Therefore, natural questions arise: can we find a length scale that is relatively larger than the average mean free path of "randomly" moving particles and yet smaller than the characteristic length of the macroscale flow system? Similarly, is there a time scale that is large enough to observe the passage of many particles through a space point yet smaller than the characteristic time scale of the flow system? The affirmative answers to these questions indicate the possibility of a new kind of continuum where spatially disconnected

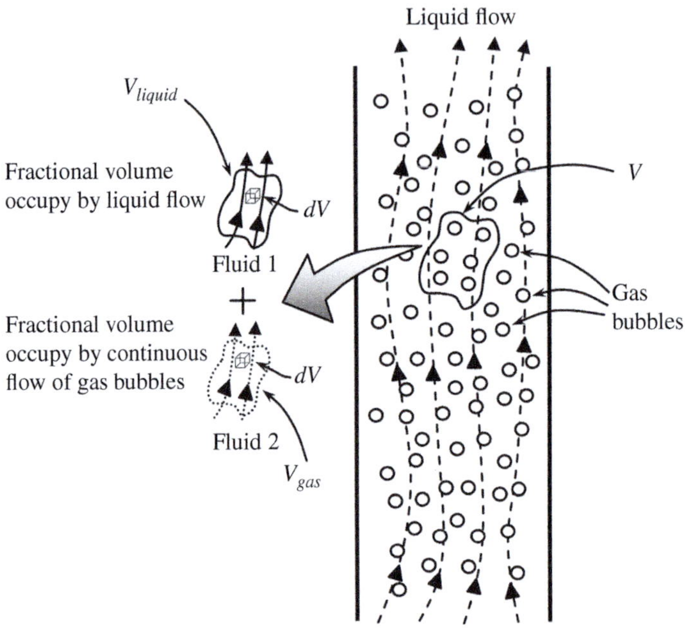

Figure 7.1 Schematic diagram shows a volume "V" covering both dispersed and continuous phases. For a sufficiently large number of dispersed phase particles, a continuum of dispersed phase can be considered. It results in the so-called "interpenetrating continua," which means both phases are present at any space point in the flow domain with their representative volume fraction. Source: Yeoh and Tu (2019)/with permission of ELSEVIER.

dispersed phase particles can also be treated like a continuum phase. This "dispersed phase continuum" provides its own set of velocity, mass, and temperature fields, which are akin to the local instant variables arising from the continuum hypothesis on the molecular level. To emphasize this point further, it can be noted that the continuous phase is already treated as a continuum, but with the usage of large enough time and length scales, even the disconnected regions of the dispersed phase can be considered as another continuum, and both continua exist simultaneously in a given volume element; see the schematic diagram provided in Figure 7.1. This idea is called interpenetrating continua in literature, and various averaging procedures are developed to find the mean variables such as velocity, mass, and temperature fields.

Eulerian time and spatial averaging are the most frequently used procedures while defining the average or mean properties such as velocity, energy, density, and concentration of each phase, especially when the dispersed phase concentration is high. These averaging procedures are also used to find the constitutive and macroscopic flow field equations in terms of mean variables. Eulerian volumetric averaging has been incorporated to model the two-phase highly dispersed flows as a mixture model by Zuber (1964a), Zuber et al. (1964), and as a two-fluid model by Delhaye (1968) and Vernier and Delhaye (1968). In these studies, both the phases are assumed to be present in a volume element at a given time instant, and linear dimension of the volume element is small in comparison to the length scale where macroscopic variables change in the macroscopic flow. Vernier and Delhaye (1968) have compared various types of Eulerian averaging methods for similarities and dissimilarities. Eulerian time averaging coupled with space averaging procedure has been used by Teletov (1945), Frankl (1953), Teletov (1957), and Diunin (1963). Eulerian time averaging alone was first proposed by Vernier and Delhaye (1968), and a detailed model was derived by Ishii (1971). The Eulerian time averaging is beneficial, especially for the case of two-phase turbulent flows, as shown by Ishii (1975) Ishii and Mishima (1984). Various shortcomings of spatial averaging procedures were pointed out by Arnold et al. (1989), and ensemble averaging has been proposed as a superior method in their work. Drew (1971) and Drew and Passman (1999) have provided a mathematically extensive framework to systematically model the multiphase flows. A concise history of the averaging methods can be found in Ishii and Hibiki (2011). For a generic Eulerian local instant variable $\phi(\mathrm{x}, t)$, the Eulerian time averaging is given by Eq. (7.4), the Eulerian volume averaging is given by Eq. (7.5), and Eulerian statistical averaging is given by Eq. (7.6). In these equations, the limit is taken to infinite, but in practice, it is taken up to sufficiently large time and length scales.

$$\overline{\phi} = \lim_{t\to\infty}\frac{1}{T}\int \phi(x,t)dt \tag{7.4}$$

$$\langle\phi\rangle_V = \lim_{V\to\infty}\frac{1}{V}\int \phi(x,t)dt \tag{7.5}$$

$$\langle\phi\rangle_E = \lim_{N\to\infty}\frac{1}{V}\sum_{n=1}^{N}\phi_n(x,t) \tag{7.6}$$

Considering a volume element "V," as shown in Figure 7.1, multiple gas bubbles (dispersed phase) occupy the space inside "V" along with the liquid (continuous phase). Because of the interpenetrating continua assumption, both phases exist simultaneously in this region. If volume-based averaging is adopted, the individual phase is said to occupy a "local space fraction," which is fractional occupancy of the volume "V" by the continuum of the kth phase at a given time instant. Similarly, if time-based averaging is adopted, then each phase occupies a space point for a fraction of averaging time interval; therefore, the continuum of the kth phase is said to have a fractional residence time.

The mass, momentum, and energy conservation equations for each phase are similar to the single-phase systems. These equations are multiplied by a phase indicator function that tracks the location of the kth phase and subsequently averaged over volume or time to provide the final governing equations for the interpenetrating continua of two phases. In the current discussion, the full derivation of continuity equation is shown, and the final momentum and energy balance equations are provided directly, as their derivations are exactly similar to the continuity equation.

Equation of Continuity Mass conservation for the kth phase across a volume element provides the continuity equation:

$$\frac{\partial \rho^k}{\partial t} + \nabla \cdot (\rho^k U^k) = 0 \tag{7.7}$$

where ρ^k and U^k are the density and the velocity of the kth phase, respectively. These are local instant quantities applicable at a space and time coordinate (x, t).

Phase Indicator Function The phase indicator variable χ^k, which is a field variable $\chi^k(x, t)$, is defined to find whether the space point is inside or outside the kth phase.

$$\chi^k = \begin{cases} 1 & \text{if } (x, y, z) \text{ is in } k^{th} \text{ phase at time t} \\ 0 & \text{otherwise} \end{cases} \tag{7.8}$$

Drew and Passman (1999) provide the governing equation for χ^k. It is clear from the definition of χ^k that when a space point is present in the bulk of any phase, its value is either one or zero. The material derivative of the phase indicator function is shown to be zero (Drew and Passman, 1999). If the interface of dispersed phase particle is moving with a velocity U^{int}, the material derivative of χ^k is given by Eq. (7.9).

$$\frac{\partial \chi^k}{\partial t} + U^{int} \cdot \nabla \chi^k = 0 \tag{7.9}$$

For the case of interpenetrated continua, the equation of continuity of the kth phase should be multiplied by the phase indicator function.

$$\chi^k \frac{\partial \rho^k}{\partial t} + \chi^k \nabla \cdot (\rho^k U^k) = 0 \tag{7.10}$$

The resulting Eq. (7.10) has a meaning that it provides a nonzero result only when the kth phase is present at the coordinate (x, t), and it produces a zero value otherwise. It also takes care of the situation when the coordinate (x, t) is present at the interface. Now, the first term on the left-hand side of Eq. (7.10) can be modified with the help of Eq. (7.9) and a vector identity operation as follows:

$$\chi^k \frac{\partial \rho^k}{\partial t} = \frac{\partial(\chi^k \rho^k)}{\partial t} - \rho^k \frac{\partial \chi^k}{\partial t} = \frac{\partial(\chi^k \rho^k)}{\partial t} + \rho^k U^{int} \cdot \nabla \chi^k \tag{7.11}$$

Similarly, the second term on the left-hand side of Eq. (7.10) can be modified using a vector identity as follows:

$$\chi^k \nabla \cdot (\rho^k U^k) = \nabla \cdot (\chi^k \rho^k U^k) - \rho^k U^k \cdot \nabla \chi^k \tag{7.12}$$

The resulting terms from Eqs. (7.11) and (7.12) can be inserted back into Eq. (7.10) to derive the final form of the continuity equation.

$$\frac{\partial(\chi^k \rho^k)}{\partial t} + \nabla \cdot (\chi^k \rho^k U^k) = \rho^k (U^k - U^{int}) \cdot \nabla \chi^k \tag{7.13}$$

The following averaging rules can be applied to Eq. (7.13) to find the instantaneous averaged equation of continuity.

$$\langle v_1 + v_2 \rangle = \langle v_1 \rangle + \langle v_2 \rangle \tag{7.14}$$

$$\langle \langle v_1 \rangle v_2 \rangle = \langle v_1 \rangle \langle v_2 \rangle \tag{7.15}$$

$$\left\langle \frac{\partial v}{\partial t} \right\rangle = \frac{\partial \langle v \rangle}{\partial t} \tag{7.16}$$

$$\langle \nabla v \rangle = \nabla \langle v \rangle \tag{7.17}$$

Following is the averaging procedure for Eq. (7.13).

$$\left\langle \frac{\partial(\chi^k \rho^k)}{\partial t} + \nabla \cdot (\chi^k \rho^k U^k) \right\rangle = \langle \rho^k (U^k - U^{int}) \cdot \nabla \chi^k \rangle \tag{7.18}$$

$$\left\langle \frac{\partial(\chi^k \rho^k)}{\partial t} \right\rangle + \langle \nabla \cdot (\chi^k \rho^k U^k) \rangle = \langle \rho^k (U^k - U^{int}) \cdot \nabla \chi^k \rangle \tag{7.19}$$

$$\frac{\partial \langle \chi^k \rho^k \rangle}{\partial t} + \nabla \cdot \langle \chi^k \rho^k U^k \rangle = \langle \rho^k (U^k - U^{int}) \cdot \nabla \chi^k \rangle \tag{7.20}$$

Equation (7.20) is called the instantaneous averaged equation of continuity. The term on the right-hand side of this equation is called interfacial mass source and is denoted by Γ^k. A similar analysis can be carried out for momentum and energy balance equations to find the final forms of instantaneous averaged equations of momentum Eq. (7.21) and energy.

Momentum Equation

$$\frac{\partial\langle\chi^k\rho^kU^k\rangle}{\partial t}+\nabla\cdot\langle\chi^k\rho^kU^kU^k\rangle=-\nabla\langle\chi^kp^k\rangle+\nabla\cdot\langle\chi^k\tau^k\rangle+\langle\chi^k\Sigma F^{k,bf}\rangle+$$
$$\underbrace{\langle\rho^kU^k(U^k-U^{int})\cdot\nabla\chi^k+\langle p^k\rangle\langle\nabla\chi^k\rangle-\langle\tau^k\cdot\nabla\chi^k\rangle} \tag{7.21}$$

Here p^k is pressure, τ^k is the viscous stress tensor, and $\Sigma F^{k,\,bf}$ is the total body force on the kth phase. The combination of three terms shown with the underbrace is called the interfacial momentum source and is denoted by Ω^k.

Energy Equation

$$\frac{\partial\langle\chi^k\rho^kH^k\rangle}{\partial t}+\nabla\cdot\langle\chi^k\rho^kU^kH^k\rangle=-\frac{\partial\langle\chi^kp^k\rangle}{\partial t}-\nabla\cdot\langle\chi^kq^k\rangle+\langle\chi^k\rangle\langle\Sigma F^{k,bf}\cdot U^k\rangle$$
$$+\left\langle\Phi_H^{\prime k}\right\rangle+\underbrace{\langle\rho^kH^k(U^k-U^{int})\cdot\nabla\chi^k\rangle+\langle q^k\cdot\nabla\chi^k\rangle+p^k\left\langle\frac{\partial\chi^k}{\partial t}\right\rangle+\langle\Phi^{extra}\rangle} \tag{7.22}$$

Here H^k is total enthalpy per unit mass. It has contributions from specific internal energy, pressure–volume work, and specific kinetic energy.

$$H^k=e^k+\frac{p^k}{\rho^k}+\frac{1}{2}|U^k|^2 \tag{7.23}$$

q^k is heat flux due to temperature gradient, and it is given by Fourier's law:

$$q^k=-\lambda^k\nabla T \tag{7.24}$$

where λ^k is the thermal heat conductivity of the kth phase. The remaining terms like $\left\langle\Phi_H^{\prime k}\right\rangle$ and $\langle\Phi^{\text{extra}}\rangle$ are defined in Eq. (7.25) and Eq. (7.26), respectively. $\left\langle\Phi_H^{\prime k}\right\rangle$ is related to the divergence of rate of work done by viscous stress forces on the volume element. $\langle\Phi^{\text{extra}}\rangle$ is the outcome of the averaging procedure.

$$\left\langle\Phi_H^{\prime k}\right\rangle=\nabla\cdot\langle\chi^kU^k\cdot\tau^k\rangle \tag{7.25}$$

$$\langle\Phi^{extra}\rangle=-\langle U^k\cdot\tau^k\cdot\nabla\chi^k\rangle \tag{7.26}$$

The last four terms of Eq. (7.22), which are grouped by the underbrace, are known as interfacial energy sources and are denoted by Π_H^k.

The equations of mass, momentum, and energy derived in Eqs. (7.20)–(7.22) are averaged for the second time using the Favre averaging procedure. The reason to do this additional averaging is that these equations contain the averages of the multiplication of various local instant variables that are difficult to determine. However, it is relatively easier to work with the averages of each individual local instant variable. The Favre averaging method allows us to convert the average of multiplication into multiplication of averages. It is a weighted averaging where different weights, such as phase indicator function or density of the kth phase, can be used as weights. In our discussion here, we use the phase indicator function as the weight to define the Favre averaging in the following Eq. (7.27).

$$\widetilde{\phi^k} = \frac{\langle \chi^k \phi^k \rangle}{\langle \chi^k \rangle} \tag{7.27}$$

The averaged phase indicator function $\langle \chi^k \rangle$ provides information about the fractional occupancy of the volume element by the kth phase. Therefore, it can be equated to the local volume fraction α^k.

$$\langle \chi^k \rangle = \alpha^k(x, t) \tag{7.28}$$

Equations (7.27) and (7.28) can be combined to provide the Favre averaging rule.

$$\langle \chi^k \phi^k \rangle = \langle \chi^k \rangle \widetilde{\phi^k} = \alpha^k \widetilde{\phi^k} \tag{7.29}$$

The decomposition of any local instant variable ϕ^k in terms of its Favre-averaged variable $\widetilde{\phi^k}$ and a fluctuating component around the phase-weighted average $\phi^{k''}$can be given as:

$$\phi^k(\boldsymbol{x}, t) = \widetilde{\phi^k} + \phi^{k''} \tag{7.30}$$

It should be noted that the time average of $\phi^{k''}$is not zero, but rather the Favre average of $\phi^{k''}$, i.e. $\widetilde{\phi^{k''}}$is zero. Eqs. (7.20)–(7.22) are Favre averaged using Eqs. (7.27)–(7.29) and (7.31), and, the *tilde* is removed from the averaged variables. The resulting equations are called the governing equation of the two-fluid model.

$$\frac{\partial(\alpha^k \rho^k)}{\partial t} + \nabla \cdot (\alpha^k \rho^k U^k) = \Gamma'^k \tag{7.31}$$

$$\frac{\partial(\alpha^k \rho^k U^k)}{\partial t} + \nabla \cdot (\alpha^k \rho^k U^k U^k) = -\nabla(\alpha^k p^k) + \nabla \cdot (\alpha^k \tau^k) - \nabla \cdot (\alpha^k \tau^{k''}) + \alpha^k \Sigma F^{k,bf} + \Omega'^k \tag{7.32}$$

$$\frac{\partial(\alpha^k \rho^k H^k)}{\partial t} + \nabla \cdot (\alpha^k \rho^k U^k H^k) = \frac{\partial(\alpha^k p^k)}{\partial t} - \nabla \cdot (\alpha^k q^k) - \nabla \cdot \left(\alpha^k q_H^{k''}\right) + \alpha^k \Sigma F^{k,bf} \cdot U^k + \Phi_H''^k + \Pi_H'^k \tag{7.33}$$

For the set of Eqs. (7.31)–(7.33) govern the Favre-averaged variables. A few terms such as $H^k = e^k + \frac{p^k}{\rho^k} + \frac{1}{2}|U^k|^2 + \frac{1}{2}|U^{k''}|^2$contains the fluctuating terms as well, which are closed by various turbulence models.

7.2.1.2 Interface Terms

The interfacial terms, Γ', Ω', and Π' are linearized in the numerical schemes of the CFD and presented as follows:

$$\Gamma'^{k} = \sum_{l=1}^{2} (\dot{m}_{lk} - \dot{m}_{kl}) \tag{7.34}$$

where the term $\dot{m}_{lk}$ means the rate of mass transfer from the lth phase to k^{th} phase.

$$\Omega'^{k} = \sum_{l=1}^{2} (\dot{m}_{lk} U_l - \dot{m}_{kl} U_k) + p_{int}^{k} \nabla \alpha^k + F_D^k \tag{7.35}$$

$$\Pi_H'^{k} = \sum_{l=1}^{2} (\dot{m}_{lk} H^l - \dot{m}_{kl} H^k) + Q_H^{int} \tag{7.36}$$

The generic models for drag force, F_D^k, and interfacial heat source, Q_H^{int}, are represented by Eqs. (7.37)–(7.38). In practice, they are modeled as algebraic functions/correlations of flow parameters.

$$F_D^{k,drag} = \sum_{l=1}^{N} B_{kl} (U^l - U^k) \tag{7.37}$$

$$Q_H^{int} = \sum_{l=1}^{N} C_{kl} (T^l - T^k) \tag{7.38}$$

7.2.2 Population Balance Model

Population balance models govern the time and phase space evolution of a particle density function, $n(\xi, x, t)$. In this review, the approach of Ramkrishna (2000) has been adopted to explain these models. The phase space comprised internal and external coordinates. The internal coordinates such as particle diameter, surface area, volume, and composition are denoted by ξ, and the external coordinates are the three spatial coordinates of a location in three-dimensional space and are denoted by the position vector x. The volume element comprised by the internal coordinates is V_ξ. This should not be understood as a physical volume in three-dimensional space. It is rather an artificial construct. Conversely, the volume element comprised by the external coordinates, V_x, is a physical volume. The total accumulation of the number of particles in the phase volume between a time interval $[0, t]$ is given by:

$$n_{total} = \int_0^t \int_{v_x} \int_{v_\xi} n(\xi, x, t) dV_\xi dV_x dt \tag{7.39}$$

The particle density function, $n(\xi, x, t)$, is assumed to have sufficient smoothness such that any required derivative is well-defined. Further, the velocity vectors can be defined in terms of both the internal and external coordinates. The rate of change of external coordinates provides the velocity in external coordinates, $v_x(\xi, x, t)$, similarly, the rate of change of internal coordinates provides the velocity in internal coordinates, $v_\xi(\xi, x, t)$. The particle fluxes (number of particles per unit area per unit time) in external and internal coordinates are given by $n(\xi, x, t)v_x(\xi, x, t)$ and $n(\xi, x, t)v_\xi(\xi, x, t)$, respectively. The final governing equation for the particle density function can be found by conserving the number of particles across a volume element present in the phase space.

$$\frac{\partial n(\xi,x,t)}{\partial t} + \underbrace{\nabla_x \cdot \left(v_x\, n(\xi,x,t)\right)}_{\text{div-in external coord-}} + \underbrace{\nabla_\xi \cdot (v_\xi\, n(\xi,x,t))}_{\substack{\text{div-in internal coord} \\ \text{or "growth term"}}} = \underbrace{(B_{br} - D_{br} + B_c - D_c)}_{S_n(\xi,x,t)} \tag{7.40}$$

Equation (7.40) is called the population balance equation. The first term on the left-hand side gives the time rate of change of $n(\xi, x, t)$ at a given space point and the internal coordinate. The second and third terms give the change of $n(\xi, x, t)$ due to advection in external coordinates and internal coordinates, respectively. It should be noted that the second term is a divergence operation in external coordinates whereas the third term is a divergence operation in internal coordinates. The third term is also called the growth term in the literature. The terms on the right-hand side of this equation jointly provide the net rate of generation of particles, $S_n(\xi, x, t)$ per unit phase volume.

The general population balance equation for the nonreactive isothermal flow cases is traditionally written in terms of a single internal coordinate; see Eqs. (7.41) and (7.42). Usually, volume, V_p, or particle diameter, d_p, are the preferred choices because the rest of the internal coordinates can be derived from either of these two variables.

$$\frac{\partial n(V_p,x,t)}{\partial t} + \nabla_x \cdot (v_x\, n(V_p,x,t)) + \frac{\partial(\dot{V}_p\, n(V,x,t))}{\partial V} = S_n(V_p,x,t) \tag{7.41}$$

$$\frac{\partial n(d_p,x,t)}{\partial t} + \nabla_x \cdot (v_x\, n(d_p,x,t)) + \frac{\partial(\dot{d}_p\, n(d_p,x,t))}{\partial d_p} = S_n(d_p,x,t) \tag{7.42}$$

In Eqs. (7.41) and (7.42), $\dot{V}_p$ and $\dot{d}_p$ are the time rate of change of volume and diameter of dispersed phase particles, respectively. In the case of gas–liquid flow, sometimes the compressibility effects are of importance, then the mass is chosen as the internal coordinate because it remains unchanged under the pressure changes. The population balance equation in terms of mass as the only internal coordinate is provided in Eq. (7.43), where $\dot{m}_p$ represents the time rate of change of mass of the dispersed phase particles.

$$\frac{\partial n(m_p, x, t)}{\partial t} + \nabla_x \cdot (v_x\, n(m_p, x, t)) + \frac{\partial(\dot{m}_p\, n(m_p, x, t))}{\partial m} = S_n(m_p, x, t) \tag{7.43}$$

Even after consideration of only one internal variable, the population balance Eqs. (7.40)–(7.43) are far too detailed, and it is difficult to incorporate them in CFD. A practical approach is to transform the population balance equation in terms of moments of the particle density function. For the internal coordinate ξ, the kth moment is given by the following expression:

$$m_k = \int_0^{\infty} \xi^k n(\xi, x, t)\, d\xi \tag{7.44}$$

Let particle diameter, d_p, be the internal coordinate, then the initial four moments of $n(\xi, x, t)$ provide the following physical variables:

$$m_0(x,t) = \int_0^{\infty} n(d_p, x, t) dd_p = N(x,t) \tag{7.45}$$

$$m_1(x,t) = \int_0^{\infty} d_p\, n(d_p, x, t) dd_p = D_p(x,t) \tag{7.46a}$$

$$\frac{m_1(\boldsymbol{x},t)}{m_0(\boldsymbol{x},t)} = \frac{\int_0^{\infty} d_p\, n(d_p, \boldsymbol{x}, t) dd_p}{N(\boldsymbol{x},t)} = \overline{D_p}(\boldsymbol{x},t) \tag{7.46b}$$

$$m_2(x,t) = \int_0^{\infty} d_p^2\, n(d_p, x, t) dd_p = \overline{a}(x,t) \tag{7.47}$$

$$m_3(x,t) = \int_0^{\infty} d_p^3\, n(d_p, x, t) dd_p = \alpha(x,t) \tag{7.48}$$

It should be noted that all the moments defined in Eqs. (7.45), (7.46a), (7.47), and (7.48) are densities of different kinds defined with respect to the physical volume rather than phase space volume. The 0th moment, m_0, in Eq. (7.45) defines the number density (with respect to physical volume) of particles of all possible diameters. The average diameter of particles is defined by the ratio of 1st and 0th moments, $\frac{m_1}{m_0}$, see Eq. 7.46(b). The 2nd moment, m_2, in Eq. (7.47) provides interfacial area concentration, i.e. total interfacial area per unit volume. Similarly, the 3rd moment, m_3, defines the total volume of particles per unit volume, i.e. the volume fraction of dispersed phase, α. If there is no interfacial mass transport between the particles and the surrounding fluid, then the $\dot{d}_p$ term is zero, which subsequently eliminates the growth term from Eq. (7.42). In this case,

Eq. (7.42) can be integrated with respect to the internal coordinate. The result is given by Eq. (7.49).

$$\int_0^\infty \frac{\partial n(d_p, x, t)}{\partial t} dd_p + \int_0^\infty \nabla_{\boldsymbol{x}} \cdot (v_x\, n(d_p, x, t)) dd_p = \int_0^\infty S_n(d_p, x, t) dd_p \tag{7.49}$$

Considering the proper smoothness of all the variables, the integral and differential operators can be commuted, as shown in Eq. (7.50). After the integration operation, the resulting equation in terms of 0th moment, i.e. $N(x, t)$, is given in Eq. (7.51).

$$\frac{\partial}{\partial t}\int_0^\infty n(d_p, x, t) dd_p + \nabla_x \cdot \left(\int_0^\infty v_x(d_p, x, t)\, n\left(d_p, x, t\right) dd_p\right) = \int_0^\infty S_n(d_p, x, t) dd_p \tag{7.50}$$

$$\frac{\partial N(x, t)}{\partial t} + \nabla_x \cdot (v_{pN}(x, t)\, N(x, t)) = S_N(x, t) \tag{7.51}$$

where $v_{pN}(x, t)$ is the number density weighted average velocity of the particles; its definition is provided in Eq. (7.52). The rate of generation per unit of physical volume is given by $S_N(x, t)$, which is further defined in Eq. (7.53).

$$v_{pN}(x, t) = \frac{\int_0^\infty v_x(d_p, x, t) n(d_p, x, t) dd_p}{\int_0^\infty n(d_p, x, t) dd_p} = \frac{\int_0^\infty v_x(d_p, x, t) n(d_p, x, t) dd_p}{N(x, t)} \tag{7.52}$$

$$S_N(x, t) = \int_0^\infty S_n(d_p, x, t) dd_p \tag{7.53}$$

7.2.2.1 Modeling of the Generation Term, S_n

The terms of $S_n(\xi, x, t)$ are explained with the help of a single internal coordinate, ξ, in the following paragraphs, but the discussion can easily be extended to the case of multiple internal coordinates, ξ, as well. We shall refer to the chosen internal coordinate, ξ, as the dispersed phase particle diameter; however, it can be chosen as shape or surface area, etc. The first term on the right-hand side of Eq. (7.40) is B_{br}, which is known as rate of birth due to breakage; its expression is given in Eq. (7.54).

$$B_{br}(\xi, x, t) = \int_{V_x}\int_\xi^\infty v(\xi', x', t) b(\xi', x', t) P(\xi, x \mid \xi', x') n(\xi', x', t) d\xi' dV_x \tag{7.54}$$

$v(\xi')$ provides the average number of particles produced when a particle of size ξ' breaks. $b(\xi', x', t)$ is breakage frequency and has a unit of reciprocal time. It depends on physical conditions, such as turbulent mixing. Higher turbulence generally leads to larger values of $b(\xi', x', t)$. $P(\xi, x \mid \xi', x')$ is daughter size distribution function; it provides the diameter distribution of daughter particles when a parent particle of diameter ξ' breaks into smaller pieces. It can also be understood as a probability density of the birth of particle having diameter ξ, when a particle of diameter ξ' breaks into smaller pieces. The last term in the integration is $n(\xi', x', t)$ which provides the number density of particles having size ξ'. Note that the limits of integration with respect to ξ' have been kept as $[\xi, \infty)$ because any particle having a diameter smaller than ξ cannot give birth to the particles of diameter ξ after the breakage. The second term on the right-hand side of Eq. (7.40) is D_{br}, which is known as death due to breakage. Eq. (7.55) provides the number of particles having diameter ξ dying per unit time because of breakage. The terms $b(\xi', x', t)$ and $n(\xi, x, t)$ are already defined in the previous paragraph.

$$D_{br}(\xi, x, t) = b(\xi, x, t)n(\xi, x, t) \tag{7.55}$$

The third term on the right-hand side of Eq. (7.40) is B_c, which is known as birth due to coalescence. Two or more particles of diameter less than ξ can combine to form a particle of diameter ξ. It should be noted that in the event of coalescence, the diameters of the two particles cannot be added to find the diameter of the new particle, instead, in this process, volumes of coalescing particles can be added to find the volume of the new particle. The expression for B_c is given in Eq. (7.56).

$$B_c(\xi, x, t) = \int_{V_x} \int_0^{\xi} \frac{1}{\delta} a(\tilde{\xi}, \tilde{x}; \xi', x', t) n(\tilde{\xi}, \tilde{x}, t) n(\xi', x, t) d\xi' dV_x \tag{7.56}$$

In Eq. (7.56), the function $a(\tilde{\xi}, \tilde{x}; \xi', x', t)$ denotes a coalescence frequency, which defines the fraction of pairs of particles having phase coordinates $(\tilde{\xi}, \tilde{x})$ and (ξ', x') coalescing per unit time. $\frac{1}{\delta}$ is a correction factor that eliminates the repeated counting of pairing between two coalescing particles. The particle density functions $n(\tilde{\xi}, \tilde{x}, t)$ and $n(\xi', x, t)$ include all those particles having phase coordinates $(\tilde{\xi}, \tilde{x})$ and (ξ', x') that can exhibit pairwise coalescence. Note that the integration limits for the internal coordinate are taken to be $[0, \xi]$ because the sizes of coalescing particles should be smaller than ξ.

The fourth term on the right-hand side of Eq. (7.40) is D_C, which is known as death due to coalescence. Its expression is given in Eq. (7.57). All the terms have been explained in the previous paragraph. It should be noted that one type of coalescing particle is given by $n(\xi, x, t)$, which means that these are the particles of diameter ξ, and when they coalesce with the other set of particles denoted by n(ξ', x, t), the resulting particle becomes larger than the diameter ξ. That also justifies the integration limit of internal variables in $(0, \infty)$ interval.

$$D_c(\xi, \boldsymbol{x}, t) = \int_{V_x} \int_0^\infty a(\tilde{\xi}, \tilde{\boldsymbol{x}}; \xi', \boldsymbol{x}', t) n(\xi, \boldsymbol{x}, t) n(\xi', \boldsymbol{x}, t) d\xi' dV_x \tag{7.57}$$

7.2.2.2 Implementation of Population Balance Model

The population balance Eq. (7.40) contains derivatives with respect to both internal and external coordinates. Additionally, the birth and death rate terms given in Eqs. (7.54)–(7.57) are integral terms. It makes the PBE a complex integro-differential equation. Solving the PBE with a typical two-fluid model requires additional constitutive equations, that is, to represent the birth and death rate expressions in terms of the existing unknown variables in order to close the system of equations. Further complexity arises because of the growth terms (represented by divergence with respect to internal coordinates in PBE). A class-based approach has been proposed by Prince and Blanch (1990) and Luo and Svendsen (1996) in which the requirement of continuous nature of particle density function in the internal coordinates has been relaxed. Instead, it is assumed that in the flow domain, there exists a finite number of classes/bins of the chosen internal coordinate. Each of these bins has its corresponding number density function, which depends only on the external coordinates. Another computationally less expensive method is called the method of moments (McGraw, 1997; Selma et al., 2010; Buffo et al., 2013).

7.2.2.3 Example of Coupling between Flow Equations and Population Balance Equations

It has been established in Section 7.2.1.2 that the continuous and dispersed phases interact via interfacial terms, which are further dependent on the particle size distribution. A schematic diagram is given in Figure 7.2 to illustrate the role of population balance model.

As an example, the flow of gas bubbles (dispersed phase) in a pool of a liquid (continuous phase) can be considered. Let us assume there is no mass and thermal energy exchange happening between the gas and the liquid via interface and only momentum interaction happens via the interfacial drag force term. A very elementary model for drag is given by Schiller and Nauman (1933) and represented here in Eq. 7.58. The quantities with a subscript "c" are related to continuous phase, and those with a subscript "d" are related to the dispersed phase, i.e. bubbles. Clearly, the drag coefficient,C_D,

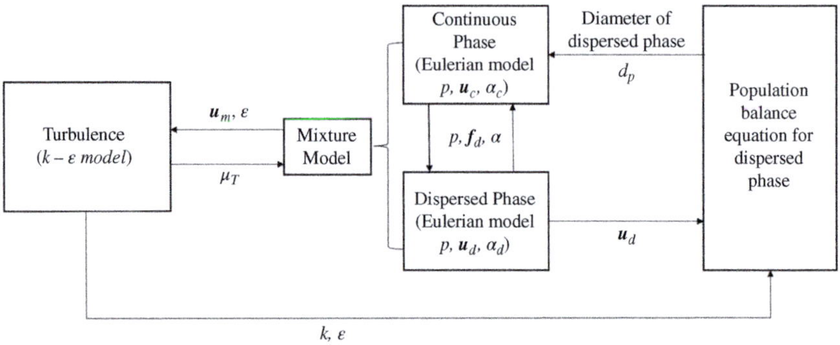

Figure 7.2 Coupling of various models in typical multiphase flow problem.

is dependent on the diameter of bubbles.

$$F_d = \frac{3\alpha_c \alpha_d C_D \rho_c (\boldsymbol{u_c} - \boldsymbol{u_d})|\boldsymbol{u_c} - \boldsymbol{u_d}|}{4d_p} \tag{7.58a}$$

$$Re = \frac{\rho_c d_p |\boldsymbol{u_c} - \boldsymbol{u_d}|}{\mu_c} \tag{7.58b}$$

$$C_D = \begin{cases} \frac{24}{Re}(1 + 0.15Re^{0.687}) & if\, Re < 1000 \\ 0.44 & otherwise \end{cases} \tag{7.58c}$$

For a monodispersed bubble condition, the bubbles in every region of the flow domain encounter the same drag force. In reality, the bubbles can break down into smaller bubbles or they can coalesce into larger bubbles during the flow. Hence, the monodisperse assumption cannot adjust for the varying drag force experienced by bubbles of varying diameters in different regions of the flow domain. Therefore, a population balance model is needed that can account for the changing diameters. The population balance model in turn depends on the flow because of various breakage and coalescence frequency kernels present in the generation term.

The schematic diagram in Figure 7.2 describes that the liquid and gas phases both are modeled by the Euler–Euler model or two-fluid model; therefore, each phase has its own set of equations with pressure, velocity, and volume fraction fields as the unknown variables. The two phases interact with each other via the common pressure field, the drag force expression, and the volume fraction field. The population balance models take the velocity input from the dispersed phase flow model and a turbulent viscosity from the turbulent model and provide the diameter field, i.e. different values of the diameter at different locations. This information is further used by the continuous phase flow model to exert a drag force field on the dispersed phase. In this way, a more realistic prediction of the flow condition can be made.

7.2.3 Turbulence Model

It is well established that a fluid flow switches its nature from being laminar to turbulent when the Reynolds number (Re) exceeds a critical value. Considering the flow in the cylindrical pipe case, the parabolic velocity profile is a valid solution of Navier–Stokes equations for all Re values. At high values ($Re > 2100$), though, this solution becomes unstable, and small deviations in the initial conditions/at the boundaries can grow in time/space to make the velocity profile differ from the parabolic solution. The ever-present disturbances continuously add unmeasured random errors (such as fluctuations in inlet velocity) at the boundaries, thus producing a fluctuating velocity profile and making it very hard to reproduce the results. This situation requires a statistical treatment of the problem. It has been found that even though the instantaneous velocity profile fluctuates randomly in space and time, its average values over sufficiently large space or time intervals are stable and well-defined. Similar to the velocity field, other local instant variables such as temperature and concentration fields also have a fluctuating nature. This explanation of transition from laminar to turbulent is very simplistic, and a detailed treatment of this topic can be found in (Sengupta, 2021). The most important characteristic of the turbulence from the engineering viewpoint is the enhanced mixing transport of various quantities because of the presence of eddies. In the next paragraphs, the transport due to molecular diffusion and turbulent fluctuating velocities has been compared.

Molecular diffusion processes transport the mass, momentum, and energy because of fluctuating molecular velocities. On the system size scale, the fluid may be present in the state of rest, but its molecules always move with isotropic fluctuating velocities. Even though the fluctuations in molecular velocities are isotropic, i.e. same in all the directions, there can still be a net transport in a particular direction because of the presence of spatial variations in mass, momentum, and energy densities in that direction. Therefore, mass transport can happen via molecular diffusion because of spatial variation of mass density; energy transport can happen via molecular diffusion because of spatial variation of energy density; and similarly the momentum transfer can happen via molecular processes because of spatial variation in momentum density. This is the reason that diffusion-driven transports happen in the (negative) direction of gradient of the density of underlying quantity.

Convective transport happens due to fluid velocity, and thus, the direction of transport is the same as the fluid velocity direction. The turbulence makes the flow velocities fluctuate in space and time. Emphasis should be placed on "fluctuating flow velocities" that are caused by turbulence in contrast with fluctuations in molecular velocities, which are always present. In the case of a statistically stationary turbulent flow, the instantaneous fluid velocity, $v(x, t)$, can be written as a summation of a mean velocity and a fluctuation around the mean.

$$v(x,t) = \overline{v(x)} + v'(x,t) \tag{7.59}$$

In Eq. (7.59), $\overline{v(x)}$ represents the mean velocity and $v'(x, t)$ represents the fluctuating velocity around the mean value. Similarly, other local instant variables such as temperature field, $T(x, t)$, and concentration field, $c(x, t)$, can be defined in Eqs. (7.60) and (7.61), respectively.

$$T(x,t) = \overline{T(x)} + T'(x,t) \tag{7.60}$$

$$c(x,t) = \overline{c(x)} + c'(x,t) \tag{7.61}$$

The terms with an over bar represent the time-averaged terms or mean quantities, and terms with a "prime" symbol represent the fluctuations around the mean values. The time-averaged values of fluctuating terms are zero by definition. The convective molar and heat fluxes can be written by multiplying the molar density and heat density with the normal velocity. In Cartesian coordinates, the convective fluxes for the x-direction can be written as follows:

$$J_x = c(x,t)\, v_x(x,t) = \left(\overline{c(x)} + c'(x,t)\right)\left(\overline{v_x(x)} + v_x'(x,t)\right) = \overline{J_x} + J_x'(x,t) \tag{7.62}$$

$$q_x = \rho c_p T(x,t) v_x(x,t) = \rho c_p \left(\overline{T(x)} + T'(x,t)\right)\left(\overline{v_x(x)} + v_x'(x,t)\right) = \overline{q_x} + q_x'(x,t) \tag{7.63}$$

The time-averaged fluxes can be found by averaging the Eqs. (7.62) and (7.63) with respect to sufficiently large time interval. The rules of the averaging are similar to Eqs. (7.14)–(7.17).

$$\begin{aligned}\overline{J_x} &= \overline{c(x,t)\, v_x(x,t)} = \overline{\overline{c(x)}\,\overline{v_x(x)} + \overline{c(x)}\, v_x'(x,t) + c'(x,t)\overline{v_x(x)} + c'(x,t) v_x'(x,t)}\\ &= \overline{c(x)}\,\overline{v_x(x)} + \overline{c'(x,t) v_x'(x,t)}\end{aligned} \tag{7.64}$$

$$\begin{aligned}\overline{q_x} &= \overline{\rho c_p T(x,t)\, v_x(x,t)}\\ &= \overline{\rho c_p \overline{T(x)}\,\overline{v_x(x)} + \rho c_p \overline{T(x)}\, v_x'(x,t) + \rho c_p T'(x,t)\overline{v_x(x)} + \rho c_p T'(x,t) v_x'(x,t)}\\ &= \rho c_p \overline{T(x)}\,\overline{v_x(x)} + \rho c_p \overline{T'(x,t) v_x'(x,t)}\end{aligned} \tag{7.65}$$

The average convective molar flux in x-direction, $\overline{J_x}$, given in Eq. (7.64) has two constituents. The first part, $\overline{c(x)}\ \overline{v_x(x)}$, is called molar flux due to average fluid velocity, and the second part, $\overline{c'(x,t)v_x'(x,t)}$, is called the Reynolds flux associated with mole (mass) transport.

$$J_x^R = \overline{c'(x,t)v_x'(x,t)} \tag{7.66}$$

The Reynolds flux, J_x^R, appears because of the averaging procedure, where the average of the multiplication of two fluctuating quantities, i.e. $\overline{c'(x,t)v_x'(x,t)}$ cannot be zero unless they are completely uncorrelated. In turbulent flows, however, there is always a correlation between the fluctuating concentration and fluctuating velocity because of the eddies. These eddies make different fluid parcels to move in a correlated fashion that carries with them the heat and mass. Similarly, the averaging procedure on local instant heat and momentum fluxes provides Reynolds fluxes for the heat and momentum, which are given in Eqs. (7.67) and (7.68).

$$q_x^R = \rho c_p \overline{T'(x,t)v_x'(x,t)} \tag{7.67}$$

$$\tau_{xy}^R = -\rho v_x' v_y' \tag{7.68}$$

Eqs. (7.65), (7.67), and (7.68) provide Reynolds fluxes for the mass (mole), heat, and momentum transport for the x-direction only. The other components can be found using the same averaging procedure given in Eqs. (7.64) and (7.65) on y and z-directional convective fluxes. These Reynolds fluxes are further approximated using the gradient-diffusion hypothesis. According to this hypothesis, the fluxes due to fluctuating flow velocities are related to the gradients of mean quantities in the same manner the fluxes due to fluctuations in the molecular velocities are related to the gradients of local instant variables.

$$J_x^R = \overline{c'(x,t)v_x'(x,t)} = -D_T \frac{\partial \bar{c}}{\partial x} \tag{7.69}$$

$$q_x^R = \rho c_p \overline{T'(x,t)v_x'(x,t)} = -\rho c_p \alpha_T \frac{\partial \overline{T}}{\partial x} \tag{7.70}$$

$$\tau_{yx}^R = -\rho v_x' v_y' = \rho \nu_T \frac{\partial \overline{v_y}}{\partial x} \tag{7.71}$$

The parameters D_T, α_T, and ν_T are called Reynolds diffusivities or turbulent diffusivities for mass, heat, and momentum transport due to fluctuating flow velocities and have $\frac{m^2}{s}$ as the unit. It should be noted that they are flow properties and not the molecular properties. The molecular diffusivities also have the same unit, $\frac{m^2}{s}$, and are generically given by $v_{rms}\ \lambda$, where v_{rms} is the root mean square velocities of the molecules and λ is the mean free path. Similarly, the Reynolds diffusivities can be generically given by $v' L_m$, where v' is the magnitude of fluctuating velocity and L_m is called the mixing length that depends on the length scale of the largest eddy present in the flow. For a typical turbulent flow case, $v' \sim 1$ cm/s and $L_m \sim 1\ cm$ and therefore, Reynolds diffusivities have an order of magnitude $10^{-4} \frac{m^2}{s}$, which is very large in comparison with the molecular diffusivity for liquids having an order of magnitude $10^{-9} \frac{m^2}{s}$. In practice, the Reynolds diffusivities can go as high as $1 \frac{m^2}{s}$ (Kumaran, 2022). As a result, the turbulence transports the mass, heat, and momentum more effectively than the molecular diffusion alone. The large values of diffusivities stem from the large mixing length or length scale of the largest eddy. The equations governing the transport of mass, momentum, and energy get modified when they are derived with the time-averaged flux terms. For example, the Navier–Stokes equations that govern the transport of momentum, start to have nine additional components of Reynolds momentum flux or Reynolds stress tensor, $-\rho \overline{v_i' v_j'}$, where i, $j = 1, 2, 3\ or\ x, y, z$ are indices to denote the components along different coordinate directions. The time-averaged form of the momentum equation was first derived by Reynolds and is famously known as the Reynolds-averaged Navier–Stokes equation or RANS; see Eq. (7.72).

$$\frac{\partial(\rho \overline{v})}{\partial t} + \nabla \cdot (\rho \overline{v}\,\overline{v}) = -\nabla \overline{p} + \nabla \cdot [\mu(\nabla \overline{v} + \nabla \overline{v}^T)] - \nabla \left(\frac{2}{3}\mu(\nabla \cdot \overline{v})\right) + \rho g - \nabla \cdot (\rho \overline{v' v'}) \tag{7.72}$$

There are a total of 10 independent variables in Eq. (7.72), comprising four time-averaged variables (time-averaged pressure and three components of time-averaged velocities) and six components of the Reynolds stress tensor (actually there are nine such components, but because of the symmetric nature of the Reynolds stress tensor, only six of them are independent). There is a need for ten equations to close the problem. But, we have only three equations written jointly in vector form in Eq. (7.72). The fourth equation comes from the time-averaged continuity equation (equation of mass conservation). For an incompressible flow, the time-averaged continuity equation can be found in Eq. (7.73).

$$\nabla \cdot \overline{v} = 0 \tag{7.73}$$

In the procedure of deriving RANS, the number of equations remains the same, but the number of unknown variables is increased by six; hence, the system of equations to govern the turbulent fluid flow is not closed, and there is a need to find six additional equations. One can proceed further by deriving the mechanical energy equation by multiplying the fluctuating velocity with Navier–Stokes equations and subsequently performing the time averaging procedure only to find 22 additional unknown terms appearing in the resulting equations (Wilcox, 2006). Thus, the problem is not closed once again. This is known as the closure problem in turbulence. The turbulence modeling exercise is generally about relating the fluctuating terms, such as the Reynolds stress tensor, in terms of average/mean flow properties/variables. The earliest approach to model the Reynolds stress tensor was proposed by Boussinesq, wherein the Reynolds stress term was assumed to follow the same mathematical structure as that of molecular diffusion of momentum in the presence of a strain rate. This is known as Boussinesq approximation (eddy viscosity hypothesis) in the literature, which is more of a hypothesis than the approximation.

$$-\rho\,\overline{v'v'} = 2\mu_T\left(\overline{S} - \frac{1}{3}(\nabla\cdot v)I\right) - \frac{2}{3}\rho kI \tag{7.74}$$

where μ_T is called turbulent viscosity or eddy viscosity, $\overline{\mathrm{S}} = \frac{1}{2}\left(\nabla\overline{\mathrm{v}} + \nabla\overline{\mathrm{v}}^{\mathrm{T}}\right)$ is time-averaged (mean) strain rate tensor, $\mathrm{k} = \frac{1}{2}\left(\overline{\mathrm{v}'_{\mathrm{x}}\mathrm{v}'_{\mathrm{x}}} + \overline{\mathrm{v}'_{\mathrm{y}}\mathrm{v}'_{\mathrm{y}}} + \overline{\mathrm{v}'_{\mathrm{z}}\mathrm{v}'_{\mathrm{z}}}\right)$ is the turbulent kinetic energy per unit mass present in the flow, and I is the identity tensor. For an incompressible turbulent flow, Eqs. (7.73) and (7.74) can be inserted in RANS Eq. (7.72).

$$\frac{\partial(\rho\overline{v})}{\partial t} + \nabla\cdot(\rho\overline{v}\,\overline{v}) = -\nabla\overline{p} + \nabla\cdot[\mu(\nabla\overline{v} + \nabla\overline{v}^T)] + \rho g + \nabla\cdot\left(\mu_T\left((\nabla\overline{v} + \nabla\overline{v}^T) - \frac{2}{3}\rho kI\right)\right) \tag{7.75}$$

$$\frac{\partial(\rho\overline{v})}{\partial t} + \nabla\cdot(\rho\overline{v}\,\overline{v}) = -\nabla\left(\overline{p} + \frac{2}{3}\rho k\right) + \nabla\cdot[(\mu + \mu_T)(\nabla\overline{v} + \nabla\overline{v}^T)] + \rho g \tag{7.76}$$

The term $\left(\overline{p} + \frac{2}{3}\rho k\right)$ is called modified mean pressure. The left-hand side of Eq. (7.76) can be further simplified using the equation of continuity to provide the final form of RANS with eddy viscosity hypothesis; see Eq. (7.77):

$$\rho\left(\frac{\partial\overline{v}}{\partial t} + \overline{v}\cdot\nabla\overline{v}\right) = -\nabla\left(\overline{p} + \frac{2}{3}\rho k\right) + \nabla\cdot[(\mu + \mu_T)(\nabla\overline{v} + \nabla\overline{v}^T)] + \rho g \tag{7.77}$$

It is notable that with the eddy viscosity hypothesis, the final form of RANS becomes similar to the original Navier–Stokes equation. The other important point is that the eddy viscosity is a scalar field, i.e. it depends on the position of the space point in contrast with the molecular diffusivity, which is a material property. Multiple models have been proposed to find the eddy viscosity. The algebraic models, also known as the zero equation, are the simplest ones (Prandtl, 1925; van Driest, 1956). A class of one-equation also exists, where the velocity is found by $k^{\frac{1}{2}}$ and an equation governing the transport of turbulent kinetic energy is solved (Kolmogorov, 1942; Prandtl, 1945). The mixing lengths are taken from the well-established experiments. The most popular eddy viscosity models that are used in various open and commercial CFD codes in current times are two-equation. where two transport equations, one for turbulent kinetic energy and other for the dissipation of energy, are solved simultaneously with the RANS. These models are discussed in the following subsections.

7.2.3.1 The $k-\epsilon$ Model

The $k-\epsilon$ model is proposed by Jones and Launder (1972), but the standard $k-\epsilon$ model is due to Launder and Sharma (1974). This model relies on two transport equations, one for the specific turbulent kinetic energy, k, and the other for the specific dissipation rate, ϵ. Here, the word "specific" means "per unit mass." The value of turbulent viscosity is found using the formula given in Eq. (7.78). The transport equation for the specific turbulent kinetic energy and specific dissipation rate are given in Eqs. (7.79) and (7.80), respectively.

$$\mu_T = C_\mu \rho \frac{k^2}{\epsilon} \tag{7.78}$$

$$\rho\left(\frac{\partial k}{\partial t} + \bar{v}\cdot\nabla k\right) = \nabla\cdot\left[\left(\mu + \frac{\mu_T}{\sigma_k}\right)\nabla k\right] + P_k - \rho\epsilon \tag{7.79}$$

$$\rho\left(\frac{\partial \epsilon}{\partial t} + \bar{v}\cdot\nabla \epsilon\right) = \nabla\cdot\left[\left(\mu + \frac{\mu_T}{\sigma_\epsilon}\right)\nabla \epsilon\right] + C_{\epsilon 1}\frac{\epsilon}{k}P_k - C_{\epsilon 2}\rho\frac{\epsilon^2}{k} \tag{7.80}$$

The term P_k is a production term and is given by Eq. (7.81).

$$P_k = \mu_T\left[\nabla\bar{v} : (\nabla\bar{v} + \nabla\bar{v}^T)\right] \tag{7.81}$$

The terms C_μ, σ_k, σ_ϵ, $C_{\epsilon 1}$, and $C_{\epsilon 2}$ are called model coefficients, and they are different for different versions of $k-\epsilon$ models. Table 7.1 provides the values of the model coefficients for various models.

A modified form of the $k-\epsilon$ model was developed using the renormalization group theory (Yakhot and Orszag, 1986; Yakhot et al., 1992), and it

Table 7.1 Values of model coefficients for different $k-\epsilon$ models.

Model	C_μ	σ_k	σ_ϵ	$C_{\epsilon 1}$	$C_{\epsilon 2}$
Jones and Launder (1972)	0.09	1.0	1.3	1.55	2.0
Launder and Sharma (1974)	0.09	1.0	1.3	1.44	1.92
Launder and Spalding (1972)	0.09	1.0	1.3	1.44	1.92

is known as RNG $k-\epsilon$ model. In this model, the equations for μ_T, k, and ϵ are still given by Eqs. (7.78)–(7.80) but the model coefficients are calculated differently and given in Eq. (7.82).

$$C_\mu = 0.085, \sigma_k = 0.72, \sigma_\epsilon = 0.72, C_{\epsilon 1} = 1.42,$$

$$C_{\epsilon 2} = \tilde{C_{\epsilon 2}} + \frac{C_\mu \lambda^3 \left(1 - \frac{\lambda}{\lambda_0}\right)}{1 + \beta\lambda^3},$$

$$\lambda = \frac{k}{\epsilon}\sqrt{2\overline{S} : \overline{S}},$$

$$\beta = 0.012, \lambda_0 = 4.38 \tag{7.82}$$

The turbulent viscosity is a scalar field because it is given in terms of k and ϵ, which themselves are scalar fields. It should be noted that the no-slip condition is followed by both the mean and fluctuating velocities. In the bulk, the mean velocity follows a plug-like profile and decreases very rapidly to zero in the regions near the walls. The viscous effects dominate the inertial forces in this region, and hence the turbulent Reynolds number (Re_T) is small here.

$$Re_T = \frac{\rho k^2}{\mu\epsilon} \tag{7.83}$$

The standard $k-\epsilon$ model is modified using damping functions f_μ, f_1, and f_2 to predict the flow within this region. The modified equations are given by

$$\mu_T = f_\mu C_\mu \rho \frac{k^2}{\epsilon},$$

$$f_\mu = exp\left(\frac{-3.4}{\left(1 + \frac{Re_T}{50}\right)^2}\right) \tag{7.84}$$

$$\rho\left(\frac{\partial k}{\partial t} + \overline{v}\cdot\nabla k\right) = \nabla \cdot \left[\left(\mu + \frac{\mu_T}{\sigma_k}\right)\nabla k\right] + P_k - \rho\epsilon \tag{7.85}$$

$$\rho\left(\frac{\partial \epsilon}{\partial t} + \overline{v}\cdot\nabla \epsilon\right) = \nabla \cdot \left[\left(\mu + \frac{\mu_T}{\sigma_\epsilon}\right)\nabla \epsilon\right] + f_1 C_{\epsilon 1}\frac{\epsilon}{k} P_k - f_2 C_{\epsilon 2}\rho\frac{\epsilon^2}{k},$$

$$f_1 = 1,$$

$$f_2 = 1 - 0.3\,exp\left(-Re_T^2\right) \tag{7.86}$$

The damping functions mentioned in Eqs. (7.84) and (7.86) are provided in Launder and Sharma (1974), and they are different in other works. The damping function f_μ reduces the value of turbulent viscosity differently in different locations of near-wall regions as it is not a constant but rather a scalar field. The role of f_2 is to decrease the loss term in ϵ equation, thereby increasing the ϵ. The effect of increased ϵ is felt by the k equation via $-\rho\epsilon$ term, which tends to enhance the decay of specific turbulent kinetic energy. For large Re_T values, the damping functions approach unity, and equations of μ_T, k, and ϵ become identical to the ones that are applicable in the bulk. The damping functions are not very reliable, especially in the case of adverse pressure gradient conditions, and therefore, the $k-\epsilon$ model is not a preferred choice to study the flows where separation or/and re-attachment of boundary layers happen. Cooper et al. (1993) and Craft et al. (1993) have examined the performance of $k-\epsilon$ model along with three other turbulence models and concluded that $k-\epsilon$ model predicts the level of turbulence in near-wall region.

7.2.3.2 The $k-\omega$ Model

The first two-equation $k-\omega$ model was developed by Kolmogorov (1942). Later, another two-equation $k-\omega$ model was developed by Saffman (1970). Further efforts were made by Wilcox (1988). The term ω is regarded as a specific turbulence dissipation rate and has a unit of $(1/s)$. Its relation to the ϵ and k is given in Eq. (7.87).

$$\omega = \frac{\epsilon}{C_\mu k} \tag{7.87}$$

$$\mu_T = \rho\frac{k}{\omega} \tag{7.88}$$

$$\rho\left(\frac{\partial k}{\partial t} + \overline{v}\cdot\nabla k\right) = \nabla\cdot\left[\left(\mu + \sigma_k^*\mu_T\right)\nabla k\right] + P_k - \beta^*\rho\omega k \tag{7.89}$$

$$\rho\left(\frac{\partial \omega}{\partial t} + \overline{v}\cdot\nabla \omega\right) = \nabla\cdot[(\mu + \sigma_\omega\mu_T)\nabla\omega] + \frac{\gamma\omega}{k}P_k - \beta\rho\omega^2 \tag{7.90}$$

$$\sigma_k^* = 0.5, \sigma_\omega = 0.5$$

$$\beta^* = 0.09, \beta = \frac{3}{40},$$

$$\gamma = \frac{5}{9} \tag{7.91}$$

The production term P_k is given by Eq. (7.81). The $k-\omega$ model is sensitive to the free stream values of ω at the inlet (Menter 1992a). The wall shear stress is over-predicted, and the flow separation does not happen (Speziale et al. 1990). These problems are addressed by Menter (1992b) in $k-\omega$ *SST* model, which is a modified version of $k-\omega$ model with an additional cross-diffusion term. Wilcox also added the cross-diffusion term in 1993a and 1993b, and his most updated version of $k-\omega$ model is provided in (2006).

7.2.3.3 The $k-\omega$ *SST* Model

The near-wall effects are better modeled by $k-\omega$ model, but they are sensitive to the free stream values of ω at the inlet. The $k-\epsilon$ model, whereas is not that successful in the near-wall regions, but it does not have $k-\omega$ model-like sensitivity problem. Therefore, it is natural to search for a model that behaves like the $k-\epsilon$ model far away from the wall and behaves like the $k-\omega$ near the wall. This is addressed in $k-\omega$ *SST* (shear stress transport) model given by Menter (1994). If the expression of $\epsilon = C_\mu k\omega$ from Eq. (7.87) is inserted in the standard $k-\epsilon$ model, the resulting equation becomes Eq. (7.92).

$$\frac{\partial(\rho\omega)}{\partial t} + \nabla\cdot(\rho\bar{v}\omega) = \nabla\cdot\left[\left(\mu+\frac{\mu_T}{\sigma_k}\right)\nabla\omega\right] + \frac{\gamma\omega}{k}P_k - \beta\rho\omega^2 + \underbrace{\frac{2\sigma_{\omega 2}}{\omega}\nabla k\cdot\nabla\omega}_{cross-diffusion} \tag{7.92}$$

Equation (7.92) is ϵ transport equation of $k-\epsilon$ model but written in terms of ω now. If it is compared with the $k-\omega$ model then the only difference is the existence of an additional cross-diffusion term. Therefore, a multiplicative factor $(1-F_1)$ can be used to blend between the $k-\epsilon$ and the $k-\omega$ model. The standard $k-\omega$ *SST* model by Menter is given in the following equations:

$$\frac{\partial(\rho k)}{\partial t} + \nabla\cdot(\rho\bar{v}k) = \nabla\cdot\left[\left(\mu+\sigma_k^*\mu_T\right)\nabla k\right] + P_k - \beta^*\rho\omega k \tag{7.93}$$

$$\frac{\partial(\rho\omega)}{\partial t} + \nabla\cdot(\rho\bar{v}\omega) = \nabla\cdot\left[(\mu+\sigma_\omega\mu_T)\nabla\omega\right] + \frac{\gamma\omega}{k}P_k - \beta\rho\omega^2 + \underbrace{2(1-F_1)\frac{\sigma_{\omega 2}}{\omega}\nabla k\cdot\nabla\omega}_{cross-diffusion} \tag{7.94}$$

$$F_1 = tanh\left(arg_1^4\right),$$

$$F_1 = 1, k-\omega\ model,$$

$$F_1 = 0 \Rightarrow k-\epsilon\ model \tag{7.95}$$

$$arg_1 = min\left[max\left(\frac{\sqrt{k}}{\beta^*\omega d}, \frac{500\nu}{d^2\omega}\right), \frac{4\rho\sigma_{\omega 2}k}{CD_{k\omega}d^2}\right]$$

$$d : distance\ from\ th\ field\ point\ to\ nearest\ wall \tag{7.96}$$

$$CD_{k\omega} = max\left(2\rho\sigma_{\omega 2}\frac{1}{\omega}\nabla k \cdot \nabla\omega, 10^{-20}\right) \tag{7.97}$$

The model coefficients are also blended between $k-\epsilon$ and $k-\omega$ models via blending function F_1; see Eq. (7.98).

$$\phi = F_1\phi_\omega + (1 - F_1)\phi_\epsilon \tag{7.98}$$

In $k-\omega$ *SST* model, a viscosity limiter is used, which results in better agreement of the results from the model and the experiment.

$$\mu_T = \frac{a_1\rho k}{max(a_1\omega, \Omega F_2)} \tag{7.99}$$

Equation (7.99) provides a turbulent viscosity similar to $\frac{\rho k}{\omega}$ but the denominator is limited by another blending function, F_2. If F_2 is very small, the turbulent viscosity takes the form of $k-\omega$ model, and for the large vorticity magnitudes (Ω) or large F_2 values, the viscosity is limited or reduced.

$$\Omega = \sqrt{2\overline{w} : \overline{w}} = \sqrt{2\overline{w}_{ij}\overline{w}_{ij}},$$

$$w = \frac{1}{2}(\nabla\overline{v} - \nabla\overline{v}^T) \tag{7.100}$$

$$F_2 = tanh\left(arg_2^2\right),$$

$$arg_2 = max\left(2\frac{\sqrt{k}}{\beta^*\omega d}, \frac{500\nu}{d^2\omega}\right) \tag{7.101}$$

$$\gamma_1 = \frac{\beta_1}{\beta^*} - \frac{\sigma_{\omega 1}\kappa^2}{\sqrt{\beta^*}}, \gamma_2 = \frac{\beta_2}{\beta^*} - \frac{\sigma_{\omega 2}\kappa^2}{\sqrt{\beta^*}},$$

$$\sigma_{\omega 1} = 0.5, \sigma_{\omega 2} = 0.856,$$

$$\beta_1 = 0.075, \beta_2 = 0.0828, \beta^* = 0.09,$$

$$\kappa = 0.41, a_1 = 0.31 \tag{7.102}$$

The predictions of $k-\omega$ *SST* model are better than the simple $k-\omega$ model.

Nomenclature

Notation		SI Unit
B_{br}	Birth rate due to breakage	variable units
B_c	Birth due to coalescence	variable units
$c(x, t)$	concentration field	mol/m^3

D_{br}	Death due to breakage	variable units
D_c	Death due to coalescence	variable units
d_p	diameter of dispersed phase particle	m
D_T	Turbulent diffusivity for mass transport	m^2/s
$F^{k,bf}$	body force per unit volume on k^{th} phase	N/m^3
g	gravitational acceleration vector	$9.8\ m/s^2$
H	specific enthalpy	J/kg
H^k	specific enthalpy of kth phase	J/kg
J^R	Reynolds mass flux	$mol/(m^2.s)$
k	Turbulent kinetic energy per unit mass	J/kg
Kn	Knudsen number	unit less
ℓ	characteristic length scale of the system	m
ℓ^*	intermediate length scale of the system	m
m_k	k^{th} moment of particle number density	variable units
m_p	mass of dispersed phase particle	kg
$N(x, t)$	particle number density function with respect to physical volume	m^{-3}
$n(\xi, x, t)$	particle number density function (with respect to phase space volume)	variable units
p	pressure	Pa
p^k	pressure of kth phase	Pa
q	conductive heat flux vector	$J/(m^2.s)$
q^k	conductive heat flux vector of kth phase	$J/(m^2.s)$
q^R	Reynolds heat flux	$J/(m^2.s)$
Q_H^{int}	interfacial heat source	W/m^3
S	shear rate tensor	1/s
$S_n(\xi, x, t)$	net rate of generation of particles per unit phase volume	variable units
t	time	s
$T(x, t)$	temperature field	K
U^{int}	velocity vector of interface	m/s
U^k	velocity vector of kth phase	m/s
V_p	volume of dispersed phase particle	m^3
V_x	volume in physical space	m^3
$v_x(\xi, x, t)$	velocity in physical space	m/s

V_ξ	volume in internal coordinates	variable units
$v_\xi(\xi, x, t)$	rate of change of internal coordinates	variable units
$v(x, t)$	velocity field	m/s

Greek Nomenclature

α_T	turbulent diffusivity for heat transport	m^2/s
λ^k	thermal conductivity of kth phase	W/m·K
ν_T	turbulent diffusivity for momentum transport	m^2/s
Π_H^k	interfacial energy sources	W/m^3
ρ^k	mass density of kth phase	kg/m^3
τ^k	shear stress tensor of kth phase	Pa
τ^R	Reynolds momentum flux tensor	Pa
α^k	local volume fraction of kth phase	unit less
Γ^k	interfacial mass source	$kg/(m^3.s)$
ϵ	specific dissipation rate	J/kg
λ	average mean free path	m
μ	viscosity	Pa.s
ξ	internal coordinate vector for population balance model	variable units
ρ	mass density	kg/m^3
τ	shear stress tensor	Pa
χ^k	phase indicator function	unit less
Ω^k	interfacial momentum source	N/m^3

References

Arnold, G.S., Drew, D.A., and Lahey, R.T. Jr., (1989). Derivation of constitutive equations for interfacial force and Reynolds stress for a suspension of spheres using ensemble cell averaging. Chem. Eng. Comm. 86: 43–54.

Bird, R.B., Stewart, W.E., and Lightfoot, E.N. (2002). Transport Phenomena. New York: John Wiley and Sons, Inc.

Buffo, A., Vanni, M., Marchisio, D.L., and Fox, R.O. (2013). Multivariate quadrature-based moments methods for turbulent polydisperse gas–liquid systems. Int. J. Multiph. Flow 50: 40–57.

Cooper, D., Jackson, D.C., Launder, B.E., and Liao, G.X. (1993). Impinging jet studies for turbulence model assessment-l. flow-field experiments. Int. J. Heat Mass Transfer. 36: 2675–2684.

Craft, T., Graham, L.J., and Launder, B.E. (1993). Impinging jet studies for turbulence model assessment-II. An examination of the performance of four turbulence models. Int. J. Heat Mass Transfer. 36: 2685–2697.

Delhaye J.M. Equations of Fondamentales des Écoulements Diphasiques, Part 1 and 2. CEA-R-3429, 1968.

Delhaye, J.M. (1974). Jump conditions and entropy sources in two-phase systems. Int. J. Multiph. Flow 1: 395–409.

Diunin, A.K. (1963). *On the Mechanics of Snow Storms*. Akademii Nauk: Siberian Branch.

Drew, D.A. (1971). Averaged field equations for two-phase media. Studies Appl. Math 1: 133–166.

Drew, D.A. and Passman, S.L. (1999). *Theory of Multicomponent Fluids*. Berlin: Springer-Verlag.

van Driest, E.R. (1956). On turbulent flow near a wall. J. Aerospace Sci. 23: 1007–1011.

Frankl, F.I. (1953). On the theory of motion of sediment suspensions. Dokl. Acad. Nauk SSSR 92: 247–250.

Ishii M. Thermally induced flow instabilities in two-phase mixture in thermal equilibrium. Ph.D. thesis, Georgia Institute of Technology, 1971.

Ishii M. Thermo-fluid dynamic theory of two-phase flow. Collection de la Direction des Etudes et Researches d'Electricite de France, Eyrolles, Paris, France, 1975, 22.

Ishii, M. and Hibiki, T. (2011). *Thermo-Fluid Dynamics of Two-Phase Flow*. New York: Springer.

Ishii, M. and Mishima, K. (1984). Two-fluid model and hydrodynamic constitutive relations. Nucl. Eng. Des 82: 107–126.

Jones, W.P. and Launder, B.E. (1972). The prediction of laminarization with a two-equation model of turbulence. Int. J. Heat Mass Transfer 15: 301–314.

Kolmogorov, A.N. (1942). The equations of turbulent motion in an incompressible fluid. Akad, Nak USSR, Ser. 6: 56–58.

Kumaran, V. (2022). *Fundamentals of Transport Processes*. Cambridge, UK: Cambridge University Press.

Launder, B.E. and Sharma, B.I. (1974). Application of the energy-dissipation model of turbulence to the calculation of flow near a spinning disc. Letters. Heat Mass Transf. 1: 131–138.

Launder, B.E. and Spalding, D.B. (1972). *Mathematical Models of Turbulence*. London: Academic Press.

Luo, H. and Svendsen, H.F. (1996). Theoretical model for drop and bubble breakup in turbulent dispersions. AIChE Journal 42: 1225–1233.

McGraw, R. (1997). Description of aerosol dynamics by the quadrature method of moments. Aerosol Sci. Tech. 27: 255–265.

Menter, F.R. (1992a). Influence of freestream values on k-w turbulence model predictions. AIAA J. 30: 1657–1659.

Menter, F.R. (1992b). Improved two-equation k-w turbulence models for aerodynamic flows. NASA Tech. Memo. 103975.

Menter, F.R. (1994). Two-equation Eddy-viscosity turbulence models for engineering applications. AIAA J. 32: 1598–1605.

Pope, S.B. (2000). *Turbulent Flows*. Cambridge: Cambridge University Press.

Prandtl, L. (1925). Bericht über die Entstehung der Turbulenz. ZAMM 5: 136–139.

Prandtl, L. (1945). Uber ein neues formelsystem fur die ausgebildete turbulenz. Nachr. Akad. Wiss. Gottingen Math-Phys. K1: 6–19.

Prince, M.J. and Blanch, H.W. (1990). Bubble coalescence and break-up in air-sparged bubble columns. AIChE J. 10: 1485–1499.

Ramkrishna, D. (2000). *Population Balances: Theory and Applications to Particulate*. San Diego: Academic Press.

Saffman, P.G. (1970). A model for inhomogeneous turbulent flow. Proc. R. Soc. Lond. Ser. A 317: 417–433.

Schiller, L. and Nauman, A. (1933). A drag coefficient correlation. Zeit. Ver. Deutsch. Ing. 77: 318–320.

Selma, B., Bannari, R., and Proulx, P. (2010). Simulation of bubbly flows: comparison between direct quadrature method of moments (DQMOM) and method of classes (CM). Chem. Eng. Sci. 65: 1925–1941.

Sengupta, T.K. (2021). *Transition to Turbulence. A Dynamical System Approach to Receptivity*. Cambridge, UK: Cambridge University Press.

Speziale, C.G., Abid, R., and Anderson, E.C. (1990). A critical evaluation of two-equation :models for near wall turbulence. AIAA Paper 90: 1481.

Teletov, S.G. (1945). Fluid dynamic equations for two-phase fluids. Soviet physics doklady. Akademii Nauk SSSR 50: 99–102.

Teletov, S.G. (1957). On the problem of fluid dynamics of two-phase mixtures, i. hydrodynamic and energy equations. Bull. The Moscow University 2: 15.

Vernier, P. and Delhaye, J.M. (1968). General two-phase flow equations applied to the thermohydrodynamics of boiling nuclear reactors. Energie Primaire 4: 1.

Wilcox, D.C. (1988). Reassessment of the scale-determining equation for advanced turbulence models. AIAA J. 26: 1299–1310.

Wilcox, D.C. (1993a). A two-equation turbulence model for wall-bounded and free-shear flows. AIAA Paper 93: 2905.

Wilcox, D.C. (1993b). Comparison of two-equation turbulence models for boundary layers with pressure gradient. AIAA J. 31: 1414–1421.

Wilcox, D.C. (2006). *Turbulence Modeling for CFD*. California: DCW Industries, Inc.

Yakhot, V. and Orszag, S. (1986). Renormalization group analysis of turbulence: l. Basic Theory. J. Sci. Comput 1: 3–51.

Yakhot, V., Orszag, S.A., Thangam, S. et al. (1992). Development of turbulence models for shear flows by a double expansion technique. Physics of Fluids A 4: 1510–1520.

Yeoh, G.H. and Tu, J. (2019). *Computational Techniques for Multiphase Flows*. Cambridge, United States: Butterworth-Heinemann.

Zuber, N. (1964a). On the dispersed flow in the laminar flow regime. Chem. Eng. Sci. 19: 897–917.

Zuber, N., Staub, F.W., and Bijwaard, G. (1964). Steady state and transient void fraction in two-phase flow systems. Two-phase Flow Systems 1: GEAP 5417.

8

Cost Estimation and Energy Consumption

8.1 Introduction

The microbubbles (MBs) have been identified and utilized in several industries in past decades, and a lot has been discussed in the previous chapters. In industrial processes, adopting a new technology might be cost-intensive as well as a huge challenge to get rid of the earlier stages. A lot of exercise is required to figure out the capital and running costs to be borne by the industry to implement a new and advanced technology. Although all the industries are expected to follow the sustainability goals and environmental norms, they also worry about their profit and loss statements. Ever since the MB technology was discovered to enhance the performance of several processes, a lot of small- and pilot-scale operations have been performed in order to test the feasibility of the technology, and some of them have also tried the upscaling of the pilot processes. It is a known fact that scaling up of any laboratory-scale process reduces the %efficiency of performance while testing on the real samples. This chapter discusses the steps to identify the operating parameters and the methods to estimate the operating cost of the MB setup, which can be integrated with the existing systems in various processes. Some of the case studies applied for pilot- and industrial-scale processes utilizing MB technology have been discussed briefly considering their energy and cost estimation.

8.2 Identification of Operating Parameters

For the MB-based processes, either water treatment or others, the major factors that can affect the overall performance of the process are either the MB size or other physical properties corresponding to the process objective. In

Microbubbles: Engineering Aspects and Industrial Applications, First Edition. Snigdha Khuntia.

MB-based water treatment, the major factors that could affect the treatment efficiency are:

(a) The size of the MBs: The MB size has a prominent role in controlling the overall mass transfer of the process, where a higher rate of mass transfer is required. To achieve higher oxygen transfer rate, or ozone transfer rate, in the secondary and tertiary water treatment processes, final MBs are required. Since the size of the MBs not only helps to achieve higher gas transfer into water but also helps to float the finer suspended solids along with the gas mass transfer.
(b) While working with gas mass transfer followed by chemical reactions such as oxidation, it is important to maintain a particular range of pH to achieve a higher oxidation percentage. In the case of ozonation-based MB processes, since ozone is highly dependent on the pH of the water, therefore the overall process becomes dependent on the pH. The surface charge of the MBs is also dependent on the aqueous pH. Therefore, the regulation and the maintenance of pH in water treatment is always a primary factor. The pollutants present in the wastewater are also pH-dependent. Hence in all the preliminary and even routine processes, the role and the effect of pH are always studied.
(c) In Chapter 3, a lot has been discussed on the hydrodynamic behavior of MBs, whereas in Chapter 2, the mass transfer properties of MBs were discussed. It was reported by all the researchers that the gas flow rate controls the master answer on hydrodynamics of MBs to a great extent. Higher gas flow rate not only enhances the nature of mass transfer of gas into liquid but also increases the chance of coalescence of bubbles. Although MBs are free from coalescence, too high a gas flow rate will reduce the MB density in the column, and in the reaction medium, the collision of pollutants and the MBs will not be sufficient. Therefore, another operating parameter that is always studied in the MB-based water treatment system is the gas flow rate or superficial gas flow rate. Similar to the gas flow rate, in the continuous processes, the flow rate of water can be a controlling step in the overall water treatment process. An optimum value of gas and water flow rate is always recommended, in order to maintain a sufficient reactant contact time.
(d) To calculate the thermodynamic parameters in the reacting and non-reacting MB-based systems, temperature acts as an important operating parameter. Most of the reaction-based systems are temperature-dependent, whereas the gas scrubbing, flotation, and cleaning processes that involve MBs are more or less dependent on temperature. The rise in internal pressure during the shrinkage of

MBs also increases the local pressure, and a high-density micromobile system has shown evidence of raising the system temperature by 10 to 30° centigrade. Maintaining a uniform temperature during a certain process is an important parameter to achieve a uniform conversion or efficiency. Therefore, on an industrial scale, temperature is considered another important parameter for optimization.

(e) The microtubule-aided processes such as flotation, cleaning, mineral beneficiation, and oil recovery are also dependent on the surface charge of the bubbles. For enhanced operation, the addition of ionic and nonionic surfactants is used to alter the surface charge of the MBs. Unlike water treatment, these processes are highly dependent on the zika potential of the MBs. It is to be noted that zeta potential is also a function of the pH of water. Hence, simultaneously, the zeta potential and the pH are the important factors.

(f) Other important parameters, such as column design and diameter, are directly proportional to the hydrodynamic parameter of the MBs. The processes such as Mittal surface cleaning, beauty industry, fish culture, and related MB-based systems are independent of column specifications. On the contrary, the bubble column-based systems such as water treatment and gas scrubbing are strongly dependent on the column dimensions. Therefore, during the process parameter optimization, the column length to diameter ratio is considered as one of the operating parameters.

(g) Apart from process parameters, MB generator design-based parameters such as pump specifications, impeller design, nozzle type, sparger type, and pressure drop are some common parameters that could affect dabble size, density, and uniformity in the system.

8.3 Cost Estimation

The cost estimation of the pilot-scale and industrial-scale operation will have different levels of entities and factors. However, in an MB-based process, the major difference in the cost comes with the replacement of generic bubble generators and MB generators.

For example, in the water treatment process, instead of the spargers used in the secondary or tertiary aeration systems, MB generator will be connected. However, if the process involves the O_3-MB, additional provisions for the O_3 generators shall be provided, since O_3 needs to be generated in situ and used immediately. The general cost estimation considerations for a

MB-based system shall be considered by taking into account the following additional points.

In the cost estimation process, it is important to include the additional systems required in the O_3-MB-based systems, such as the O_2 concentrator, which is necessary equipment for continuous supply of O_2 to the O_3 generator, along with the O_3 destructor to completely destruct the spent O_3 released from the system. Due to the strict environmental norms, it is mandatory practice to destruct the unreacted O_3 before releasing it to atmosphere. Considering the simple calculations step discussed in Table 8.1, the MB-based systems should be considered as replacement over the conventional system only if the following advantages are noticed, in the industrial process.

(i) If the treatment process or desired objective has achieved a noticeable change after incorporating the MBs. At the industrial level, if the process or desired objective performance is enhanced by at least 10%, only then the investment in the MB system appear to be feasible.
(ii) Since MB systems are known to generate less sludge and have lower requirements of additional chemicals (in water treatment process), this could act as an added benefit to the process.
(iii) For the purposes that involve the generation of reactive oxygen species, the convenient method is to use MBs combined with the appropriate catalyst and hence could be used industrially (Singh et al., 2024). Other processes, such as Fenton and ultrasonication method, also require approximately the same investment and running cost as that of MB systems, but the latter has the advantage of generating less sludge.
(iv) The cost of pump requirement in water displacement and recirculation is inevitable in the industrial practices; hence, there shall not be more deference in overall cost requirement for the food processing and fishery industries to add the MB system.

The major power difference in the MBs over the conventional pumps is decided by the type of MB generator used for the process. The porous membrane-based MB system will require a compressor, whereas the pressurized system will require the impeller-based centrifugal pump (discussed in Chapter 1). The calculation of energy requirement for the MB generator is discussed in Section 8.3.

8.4 Power Measurement

The power measurement or consumption of fluid machine is measured depending on the power required for the impeller to rotate at a certain

Table 8.1 Steps involved in calculating the cost of MB system.

(A) For air-MB systems	(B) O_3-based MB system
(i) MBG type	(i) Total capital and running cost of step (A)
(ii) Power consumption of	(ii) O_3 generator
(iii) MBG	(iii) Power consumption of O_3 generator
(iv) Scale of MBG	(iv) O_2 concentrator
(v) Operation time	(v) Power consumption of O_2 concentrator
(vi) Capital cost of MBG	(vi) spent O_3 destructor
(vii) Running cost/day	(vii) Total operation time
	(viii) Overall capital and running cost for the O_3-MB system

rotation per minute. In the case of a non-gas system, the general expression for the power consumption for a Newtonian fluid is calculated by the well-known power number N_P:

$$N_P = \frac{P_o}{N^3 D_{i\ \rho L}^5} \tag{8.1}$$

where P_o is the power supplied to the impeller/agitator, N is the rotational speed of the impeller/agitator, D_i is the impeller/agitator diameter, and ρ_L is the liquid density. In agitated systems, the Reynolds number is directly proportional to the fluid flow characteristics by the following equation:

$$Re = \frac{N D_{i\ \rho L}^2}{\mu_L} \tag{8.2}$$

It has been studied by many researchers that the presence of a type of gas inside the shrinking MB can also affect the power consumption of the microbial generator. When a bubble comes into the picture in the agitated systems, the difference in densities between the gas and the liquid comes into the picture. The gas bubbles tend to reduce the density of the fluid mixture and require less energy consumption. Therefore, the power consumption equation above can be modified for a gas-based bubble system by incorporating the dimensionless aeration number (N_a) (Hensirisak et al., 2002):

$$\frac{P_G}{P_0} = N_a = \frac{f_1 \times Q}{N D^3} \tag{8.3}$$

where P_G = power supply for gassed system (W), P_0 = power supply for un-gassed system (W), f_1 = function obtained graphically, Q = volumetric gas flow rate (L/s), D = impeller/agitator diameter (cm), and N = speed of the impeller/agitator (rev s^{-1}).

The MB generator includes mostly three power-consuming devices, such as a motor, a MB delivery pump, and a recycling pump. Therefore, the power consumption of the typical MB generator is a combination of these three sections and is defined as:

$$P_{MBG} = P_{motor} + P_{pump} + P_{recycle\ pump} \tag{8.4}$$

$$P_{pump} = Current \times voltage \tag{8.5}$$

A steady-state mechanical energy balance at a macroscopic level is used to calculate the power of the MB delivery and recycling pump.

$$\Delta\frac{1}{2}\langle v^2\rangle + g\Delta h + \int_{p_1}^{p_2}\frac{1}{\rho}dp + W + \sum_i\left(\frac{1}{2}\langle v^2\rangle\frac{L}{D}f\right)_i + \sum_i\left(\frac{1}{2}\langle v^2\rangle e_v\right)_i = 0 \tag{8.6}$$

where $\langle v\rangle$ is the fluid average velocity, Δh is the height through which the fluid is lifted, L_P and D_P are the length and diameter of the pipe, f is the friction factor, e_v is loss due to fittings (frictional loss), and W is the pump shaft work. The power consumed by the air compressor to create the MB is calculated as follows:

$$H_P = 0.0154QPX \tag{8.7}$$

where H_p is the power (hp), Q is the flow rate of the fluid, P is pressure (psi), and X is a correlation factor according to Perry et al. (1984).

8.5 Experimental Method of Evaluation of the Performance of Microbubble Generators

One of the methods for the evaluation of the performance of the MB generator is the experimental route by taking the help of oxygen dissolution capacity of the machine. The factors that directly affect the performance of the MB generator are the ratio of flow rate (FR) of water–air, water renewal time (RT), and area-to-volume ratio (AVR) of water under aeration as compared to MB generator. AVR also represents the inverse of the height of water under aeration and provides prolonged time for flotation of MBs subjected to maximum oxygen transfer. The factors FR, RT, and AVR can be expressed as follows:

$$FR = \frac{Q_{H_2O}}{Q_{air}} \tag{8.8}$$

$$RT = \frac{Q}{Q_{H_2O}} \tag{8.9}$$

$$AVR = \frac{A}{Q} \tag{8.10}$$

Q_{H_2O} and Q_{air} are the flow rates of water (L/min) and air (L/min), A is the area of the upper layer of water, and Q is the volume of water, respectively. Another representative function $k_L a_{20}$ is used to indicate the dissolution rate of oxygen in water by using the standard O_2 transfer rate (SOTR) and the standard aeration efficiency (SEA) (Ochoa et al., 2022):

$$\text{SOTR} = k_L a_{20} C_{s,20} V \tag{8.11}$$

$$\text{SEA} = \frac{\text{SOTR}}{H_P} \tag{8.12}$$

$$k_L a_{20} = k_L a \theta^{(20-T)} \tag{8.13}$$

The procedure to calculate $k_L a$ has been discussed in Chapter 2. The term θ is a constant, and the recommended approximation is 1.024. H_P is the required power for the aeration system. For comparison of multiple MB generation system efficiency, the above set of equations can be used and the values of SEA and SOTR can be compared.

In a recent work reported by Jiang et al. (2022), a novel high-gravity MB generator (or HiGee) has been designed and fabricated. A new approach to calculate the energy dissipation rate (ε_D) have been developed and explained to calculate the value through experimental route. The ε_D is defined as the amount of energy lost during the turbulent flow by the viscous forces. The size of the MBs can be controlled if the controlled value of ε_D can be achieved in the device, which significantly affects the value of d_{32}. The ε_D of the HiGee MB generator was evaluated as:

$$\varepsilon_D = \frac{P_F}{\rho_L V_{MBG}} \tag{8.14}$$

P_F is the power to work on the fluid (W), and V_{MBG} is the volume of MB generator (m^3).

8.6 Estimation of Energy Consumption for Lab-Scale Microbubble-Based System

A method followed for the estimation of the electrical energy requirement for a laboratory or small-scale pilot study is calculation of the figure of merit. However, the figure of merit can be defined on two different bases (Bolton et al., 2001; Khuntia et al., 2021).

(1) E_M, electric energy per mass, which is defined as the amount of electrical energy consumed to degrade the target pollutant per unit mass.

in batch operations

$$E_M = \frac{Pt1000}{VM(c_i - c_f)}, \text{kWh/kg} \tag{8.15}$$

in continuous operations

$$E_M = \frac{P}{QM(c_i - c_f)}, \text{kWh/kg} \tag{8.16}$$

P will be the power requirement for the MB generator, t = operation time, V = volume of water to be treated (L), M = molar mass of pollutant (g/mol), Q = volumetric water flow rate (m^3/h), and c_i and c_f are the concentrations of pollutant at initial and final time (mol/L), respectively. Higher E_M values represent the lower removal efficiencies.

(2) E_{EO}, electric energy requirement per order, is the amount of energy consumed (KWh) to treat the pollutant to bring it down to an order of initial concentration.

in batch operations

$$E_{EO} = \frac{Pt1000}{V\log(c_i - c_f)}, \frac{\frac{\text{kWh}}{m^3}}{\text{order}} \tag{8.17}$$

in continuous operations

$$E_{EO} = \frac{P}{F\log(c_i - c_f)}, \frac{\frac{\text{kWh}}{m^3}}{\text{order}} \tag{8.18}$$

For O_3-MB system, the value of P will be calculated as follows:

$$P = P_{MB\ generator} + P_{O_3\ generator} + P_{O_2\ concentrator} \tag{8.19}$$

8.7 Case Studies on Application of Microbubbles in Industrial Processes

Some innovative works and their case studies for the application of MBs in some of the nonconventional applications have been summarized.

8.7.1 A Case Study on the Treatment of Highly Saline Wastewater Using Ozone Micro-Nano Bubbles (O_3-MNBs)

8.7.1.1 Background and Objectives

Xia and Hu (2019) conducted a study to explore the feasibility and effectiveness of utilizing ozone micro-nano bubbles (O_3-MNBs) for the treatment of highly saline wastewater. The study focuses on evaluating the treatment efficiency of O_3-MNBs under various pH and salinity conditions, as well as their effectiveness in removing organic contaminants from industrial wastewater and groundwater. There has been a lot of work reported on the application of O_3-MNBs in the water treatment, but not for the saline water. Therefore, specifically, this study aimed to investigate the influence of pH and salinity on the treatment efficiency of O_3-MNBs and assess their ability to remove organic contaminants from industrial wastewater and groundwater.

8.7.1.2 Methodology

(a) **Laboratory Experiments:** The researchers have used the experimental route by designing a special setup for the injection of the O_3-MNBs. The laboratory-scale experiments were initially conducted using methyl orange as a model contaminant to evaluate the treatment efficiency of O_3-MNBs under different pH and salinity conditions. The degradation of methyl orange was monitored to assess treatment effectiveness.

(b) **Treatment of Highly Saline Wastewater:** Several experiments were carried out to treat highly saline wastewater samples to evaluate the removal efficiency of organic contaminants using O_3-MNBs. The chemical oxygen demand (COD) removal rate was measured to quantify treatment effectiveness.

(c) **Remediation of Contaminated Groundwater:** Contaminated groundwater containing complex, persistent organic compounds was remediated using O_3-MNBs. The concentrations of benzene and chlorobenzene were monitored before and after treatment to assess contaminants removal.

8.7.1.3 Results

(a) **pH and Salinity Effects:** Treatment efficiency of O_3-MNBs was influenced by pH and salinity conditions. Optimal treatment occurred at pH 5, with salt accelerating methyl orange degradation. However, caution was advised due to the potential formation of chlorinated aromatic compounds under certain conditions.

(b) **Treatment of Highly Saline Wastewater:** O_3-MNBs exhibited significant efficacy in removing organic contaminants from highly saline wastewater, with a COD removal rate exceeding 63% after a 14-hour treatment period, as shown in Figure 8.1.

(c) **Remediation of Contaminated Groundwater:** Ozone treatment demonstrated remarkable effectiveness in removing persistent organic contaminants from contaminated groundwater, with substantial reductions observed in benzene and chlorobenzene concentrations after just 30 minutes of treatment.

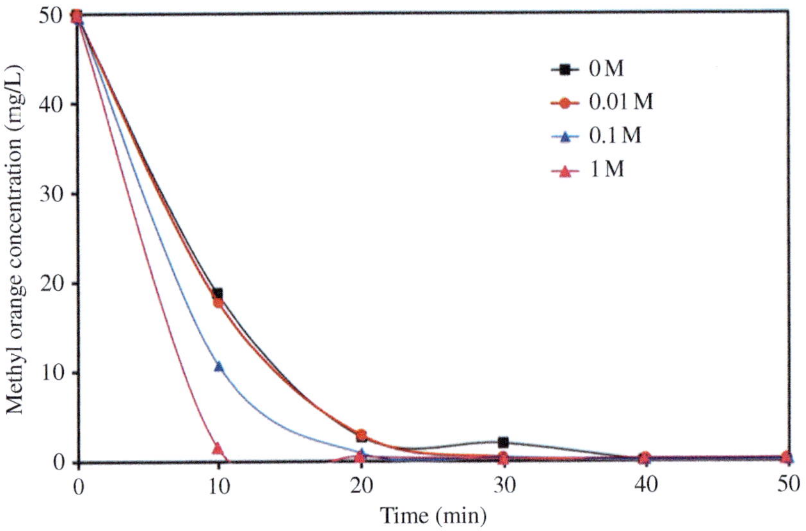

Figure 8.1 The role of salinity in the degradation of methyl orange dye using O_3-MNBs. Source: Xia and Hu (2019)/MDPI/CC BY 4.0.

8.7.1.4 Conclusions

The study concluded that O_3-MNBs represent a promising technology for the treatment of highly saline wastewater. Although there was a strict limit to maintain the pH of water and salinity level, O_3-MNBs still showed significant efficacy in removing organic contaminants from both industrial wastewater and contaminated groundwater. The findings underscore the potential of O_3-MNBs as an environmentally friendly and effective solution for wastewater treatment in saline environments. Further research and development in this area could lead to broader applications and advancements in wastewater treatment technology.

8.7.2 A Case Study on MB-Mediated Simultaneous Removal of NO and SO_2

8.7.2.1 Background and Objectives

A sustainable and rapid process was developed to address the simultaneous removal of NO and SO_2 from industrial emissions. This innovative approach utilized an MB dispersion system to facilitate the oxidation followed by absorption of NO and SO_2 in an absorption column reactor. The process aimed to achieve high removal efficiencies for both pollutants simultaneously while minimizing environmental impact. The study also includes the optimization of the process parameters and evaluates the performance of the MBs in achieving high removal efficiencies.

8.7.2.2 Methodology

(a) **Experimental Setup:** The study utilized an MB generator to create the gas mixture (NO, SO_2, and/or air) with an absorption liquid. The MBs were then introduced into an oxidation–absorption column reactor, where NO and SO_2 underwent oxidation and absorption.

(b) **Effect of Additives:** Various additives, including air, H_2O_2, and/or $Na_2S_2O_8$, were introduced into the microbubble generator (MBG) to enhance removal efficiency. The impact of additive concentrations on removal performance was investigated.

(c) **Optimization of Process Parameters:** The effects of initial pH and temperature of the absorption liquid on the simultaneous removal of NO and SO_2 were evaluated to optimize process conditions and maximize removal efficiencies.

8.7.2.3 Results

(a) **High Removal Efficiencies:** The MBs revealed remarkable performance in removing NO and SO_2, achieving removal efficiencies of 78% for NO and 94.4% for SO_2 under optimal conditions. The challenge in the removal of NO is its poor solubility in water, which was enhanced by incorporating the MBs in this process.

(b) **Effect of Additives:** The incorporation of additives such as air, H_2O_2, and $Na_2S_2O_8$ significantly improved removal performance, with complete removal of SO_2 achieved under certain conditions.

(c) **Optimized Process Parameters:** The study identified optimal conditions for NO and SO_2 removal, including pH and temperature ranges that maximized removal efficiencies without significant impact on absorbent performance.

8.7.2.4 Conclusions

The study demonstrated the effectiveness of MB-mediated process for the simultaneous removal of NO and SO_2 from industrial emissions. By leveraging MBG technology and optimizing process parameters, high removal efficiencies were achieved, contributing to environmental sustainability and air quality improvement efforts. The findings highlight the potential of this innovative approach to address multiple pollutants simultaneously, offering a promising solution for emission control in industrial settings. Further research and development could lead to broader applications of MB-mediated processes for air pollution control and environmental remediation initiatives.

8.7.3 A Case Study on Enhancing Biodiesel Production Through MB-Mediated Ozonation Process

8.7.3.1 Background and Objectives

Zimmerman and Kokoo (2018) aimed to revolutionize the pretreatment stage of biodiesel production by replacing traditional acid pretreatment with an MB-mediated reactive distillation process. This innovative approach utilizes ozone MBs to enhance the esterification process, leading to higher conversion rates and improved efficiency.

The objective of the study was to investigate the feasibility and benefits of using MB-mediated ozonation in biodiesel production, with a focus on process intensification and improved conversion rates.

8.7.3.2 Methodology

(a) **Experimental Setup:** The study conducted experiments to evaluate the efficacy of ozone MBs in the esterification process. Various alcohols were used for ozonolysis, and the resulting ester species were analyzed using GC–MS.
(b) **Comparison with Traditional Methods:** The results obtained from MB-mediated ozonolysis were compared with those from traditional acid pretreatment methods, such as ASTM D974, to assess the effectiveness and efficiency of the new approach.
(c) **Effect of Process Parameters:** The influence of process parameters, such as temperature and specific interfacial area, on the ozonolysis process was investigated to optimize the biodiesel production process.

8.7.3.3 Results

The exciting findings and results of this work have been summarized as follows:

(a) **Higher Conversion Rates:** MB-mediated ozonolysis resulted in significantly higher conversion rates compared to traditional acid pretreatment methods. The process achieved a yield exceeding 80% even without high excess methanol.
(b) **Advantages Over Acid Catalysis:** The use of ozone MBs offered several advantages over acid catalysis, including higher conversion rates, absence of catalyst residue, and elimination of additional catalyst recovery steps.
(c) **Impact of Ozonolysis:** Ozonolysis played a crucial role in breaking down carbon–carbon double bonds into aldehydes and carboxylic acids, resulting in the formation of various ester species suitable for biodiesel production.

8.7.3.4 Conclusion

The study demonstrated the effectiveness of MB-mediated ozonolysis in enhancing the pretreatment stage of biodiesel production. By leveraging ozone MBs, the process achieved higher conversion rates and improved efficiency compared to traditional methods. The findings underscore the potential of this innovative approach to revolutionize biodiesel production, offering a more sustainable and efficient alternative to conventional methods. Further research and optimization of process parameters could lead to even greater advancements in biodiesel production technology.

8.7.4 A Case Study on Innovative Methods for Fine Particle Recovery Using MBs

8.7.4.1 Background and Objectives

Parmar and Majumder (2016) investigated the technical feasibility of a novel approach for recovering fine particles using MB dispersion and flotation techniques. Traditional methods often struggle with efficient separation of fine and ultrafine particles, particularly when dealing with particles of similar sizes and charges. This study explored the potential of MBs to overcome these challenges and enhance particle recovery efficiency. The effectiveness of MBs in recovering fine particles and understanding the factors influencing their separation efficiency was evaluated.

8.7.4.2 Methodology

(a) **Experimental Setup:** The study conducted experiments to evaluate the recovery of mineral particles using MBs. Factors such as surfactant concentration, MBs and particle size, zeta potential, surface potential, and circulation velocity were varied to analyze their effects on particle recovery.

(b) **Particle Recovery:** Different types of particles, including ZnO, Al_2O_3, CuO, and SiO_2, were subjected to MB-assisted recovery. The efficiency of recovery was measured under various conditions, including different surfactants and concentrations.

(c) **Flotation Model:** A flotation model was developed to understand the underlying mechanisms of particle–MB interaction. This model incorporated collision, attachment, and stability efficiencies to calculate flotation rate constants.

8.7.4.3 Results

(a) **Surfactant Influence:** The choice and concentration of surfactants significantly affected particle recovery. CTAB demonstrated maximum recovery for ZnO and Al_2O_3 particles, while SDS and Tween-20 were more effective for CuO particles.

(b) **Effect of Zeta Potential:** The zeta potential of MBs played a crucial role in particle recovery. SDS MBs exhibited higher recovery rates due to their higher zeta potential magnitude.
(c) **Flotation Model:** The developed flotation model highlighted the importance of physicochemical properties in particle–MB interactions. Hydrophobic surface forces facilitated particle collection by MBs, enhancing separation efficiency.

8.7.4.4 Conclusion

The study demonstrated that MBs offer a promising approach to fine particle recovery, with the potential to overcome challenges associated with traditional methods. By understanding the factors influencing particle–MB interactions, such as surfactant concentration and zeta potential, efficient separation of various types of fine particles can be achieved. This innovative method holds significant promise for applications in industries requiring fine particle separation, contributing to improved efficiency and sustainability.

8.7.5 A Case Study on MB-Mediated Membrane De-fouling in Wastewater Treatment

8.7.5.1 Background and Objectives

In wastewater treatment processes, membrane fouling due to impurities and colloidal substances significantly reduces filtration efficiency. Harun and Zimmerman (2019) conducted a study to assess the effectiveness of MBs generated through fluidic oscillation in de-fouling membrane filters fouled with surface seawater. The potential of fluidic oscillator-generated MBs in removing colloidal particles deposited on membrane surfaces was the objective of the work, thereby restoring filtration performance.

8.7.5.2 Methodology

(a) **Fouling Process:** Microfiltration (MF) membranes with 1 μm pore size were used for the cartridge filters and fouled using surface seawater samples collected from the Hull coastal area. At a circulation rate of $5.8\ \mathrm{L\ min^{-1}}$, the seawater was passed to induce the deposition of the colloidal substance on the membrane surface.
(b) **Fouling Assessment:** The pressure drop at the feed channel across the membrane filters (ΔP) was recorded to monitor fouling progression. Rapid fouling was observed within the first eight hours of circulation, with fluctuations in ΔP observed over the subsequent eight hours as colloids filled the membrane pores, leading to complete fouling.

(c) **Cleaning Process:** The fouled membranes underwent de-fouling using fluidic oscillator-generated MBs. Air scouring at a rate of 1 L/min was applied for approximately one to two hours to dislodge deposited colloids and impurities from the membrane surface.
(d) **Analysis:** The cleaned membranes were analyzed using scanning electron microscopy (SEM), UV254, and an EC meter to assess the extent of MB-mediated removal of deposited colloidal particles from the membrane surfaces.

8.7.5.3 Results

The study revealed varying de-fouling rates depending on the type of MBs used:

MBs generated by a closed-vent fluidic oscillator exhibited the highest de-fouling rate of 9.53 mbar/min.

MBs generated by an opened-vent fluidic oscillator followed with a de-fouling rate of 6.22 mbar/min.

MBs generated without a fluidic oscillator demonstrated the lowest de-fouling rate of 3.41 mbar/min.

8.7.5.4 Conclusion

The findings suggest that fluidic oscillator-generated MBs are effective in removing colloidal particles deposited on membrane surfaces, thereby mitigating fouling and restoring filtration performance in wastewater treatment systems. This highlights the potential of MB-mediated membrane de-fouling as a promising strategy for improving membrane filtration efficiency in wastewater treatment plants (WWTPs).

8.7.6 A Case Study on the Separation of Low-concentration Molybdenum From Tungstate Solution

8.7.6.1 Background and Objectives

The efficient separation of lower concentrations of molybdenum fine particles from tungstate solutions has long been a technical challenge in tungsten metallurgy and the subsequent value-added application of tungsten products. Traditional methods such as solvent extraction have limitations in achieving high-efficiency separation. To address this challenge, Su et al. (2022) proposed and employed a novel strategy named MBs floating-extraction, aiming to achieve highly efficient separation of molybdenum fines from tungstate solutions. The systematic steps involved in their process is shown in Figure 8.2. The niche objective of the study was to develop a feasible and efficient method for the separation of molybdenum

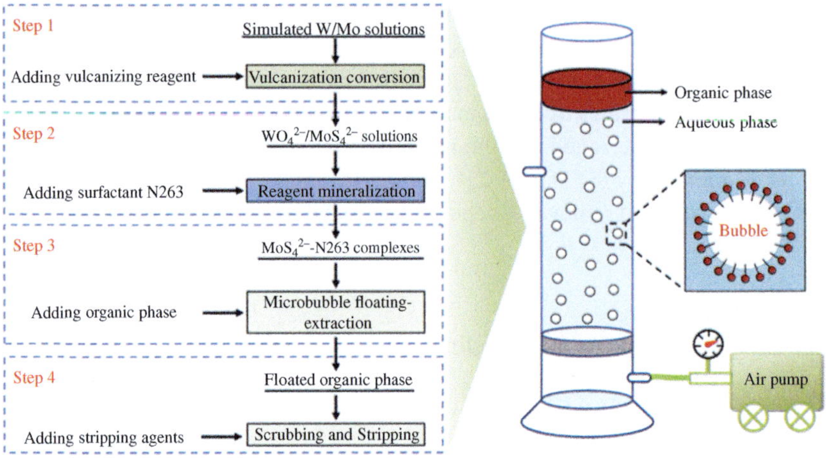

Figure 8.2 The systematic steps involved in the separation of Mo fines from the tungsten solutions using the MB-based flotation system. Source: Su et al. (2022)/American Chemical Society.

from tungstate solutions, overcoming the technical bottlenecks associated with traditional separation techniques.

8.7.6.2 Methodology

(a) **Reagent Mineralization**: The study emphasized the importance of reagent mineralization for the formation of hydrophobic complexes, $MoSO_4^{2-} - N263$, which play a crucial role in MBs floating-extraction.

(b) **MBs Floating-Extraction**: The process leveraged the higher interface mass transfer rates to achieve efficient enrichment and deep separation of $MoSO_4^{2-} - N263$ complexes. This method significantly reduced the utilization of organic solvents compared to traditional solvent extraction techniques.

(c) **Optimization of Conditions**: The optimal conditions for MBs floating-extraction were determined, including solution pH, N263 dosage, mineralization time, isooctanol concentration, flow rate, flotation time, and MB size.

8.7.6.3 Results

(a) **Efficiency Improvement:** Under optimal conditions, the efficiency of flotation of molybdenum fines increased from 98.61% to 99.14%, while the loss rate of tungsten decreased from 3.31% to 2.12% compared to solvent extraction methods.

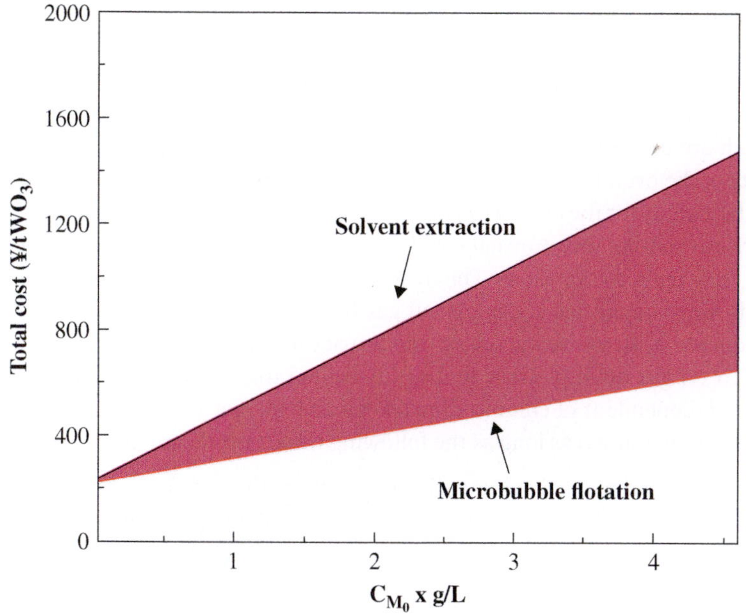

Figure 8.3 Cost estimation of the solvent extraction process and MBs flotation system and comparison. Source: Su et al. (2022)/American Chemical Society.

(b) **Cost Analysis:** The overall cost of MB-induced floating-extraction was significantly lower as compared to solvent extraction process, indicating economic advantages, as indicated in Figure 8.3.

8.7.6.4 Conclusion

The study demonstrated that MBs floating-extraction offers noteworthy technical advantages, including larger separation efficiency, a higher enrichment coefficient, lower consumption of organic solvents, and reduced economic costs. These findings suggest that MBs floating-extraction has promising application prospects in the field of separating similar metals, paving the way for improved efficiency and sustainability in tungsten metallurgy and related industries.

8.8 Scale-up Models

The scale-up models and study of the MB-based system will be similar to the conventional bubble columns. Based on the literature for the models and methodology of the scale-up of bubble columns, the following is a summary of the various acceptable criteria.

Based on the experiential and literature work, Wilkinson et al. (1992) proposed three criteria for scaling up high-pressure bubble column reactors, which are discussed below. These criteria are based on analyzing the similarity of the overall gas holdup. They had also developed a correlation based on their study for the overall gas holdup that incorporates not only their own work but also the literature data, which accounts for the effects of gas density and flow regime transitions. The hydrodynamic parameters involved in the study were quantified using overall gas holdup. Additionally, the authors conducted a literature survey on the impact of liquid height and sparger design on gas holdup. Their findings indicated that the gas holdup is virtually independent of column dimensions and sparger layout (at both low and high pressures) as long as the following three criteria are met.

(a) The recommended column diameter is more than 15 cm.
(b) The proposed ratio of column height to diameter has to be more than 5.
(c) The sparger hole diameter for generation of bubbles has to be larger than 1–2 mm.

However, as a conclusion, Wilkinson et al. (1992) proposed the use of experimental data from pilot plant bubble columns with small dimensions (aspect ratio < 5, $D < 15$ cm) or with porous plate-type spargers to estimate hydrodynamic parameters (such as gas holdup, volumetric mass transfer coefficient, and interfacial area) for large-scale industrial bubble columns, which will generally result in significant overestimation of these parameters.

A study performed by Degaleesan (1997) proposed a scale-up methodology based on the assumption that any gas–liquid or gas–slurry system would show similar hydrodynamic behavior to an air–water system if both systems have the same overall gas holdup. This procedure involves measuring the overall gas holdup in a scaled-up unit under its operating conditions based on an appropriate correlation. Then, the equivalent superficial gas velocity is calculated, which provides the same overall gas holdup in an atmospheric air–water system as in the scaled-up unit. Consequently, it was suggested that the hydrodynamics and mixing at this equivalent superficial gas velocity in an atmospheric air–water system would represent the hydrodynamics and mixing in the scaled-up unit.

Inga and Morsi (1997) developed a scale-up/scale-down methodology for bubble and slurry bubble columns based on the similarity of the relative importance of mass transfer resistance in the overall reaction resistances. They demonstrated their procedure with Fischer–Tropsch (FT) synthesis, showing that experimental results obtained in a laboratory-scale stirred tank reactor could be extrapolated to design an industrial-scale slurry bubble column. The relative importance of mass transfer resistance was

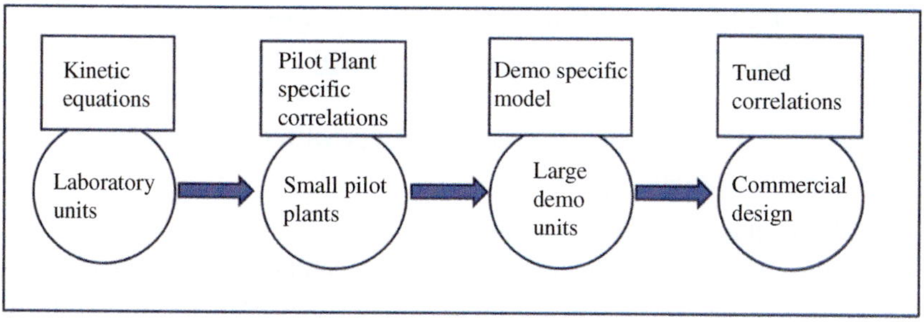

Figure 8.4 Linear scale-up process in traditional hierarchy. Source: Tarmy and Coulaloglou (1992)/with permission of ELSEVIER.

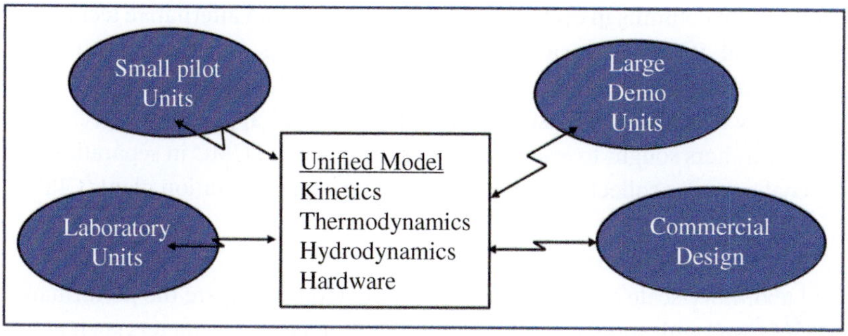

Figure 8.5 Scale-up strategy for industrial process. Source: Tarmy and Coulaloglou (1992)/with permission of ELSEVIER.

defined using a dimensionless parameter, βi, which represents the balance between the mass transfer coefficient ($k_L a$) and the rate of consumption (k_0), the pseudo-kinetic constant for a first-order reaction, as follows:

$$\beta_i = \frac{\frac{1}{k_L a}}{\frac{1}{k_L a} + \frac{1}{k_0}} \tag{8.20}$$

The status of scale-up of bubble column reactors in industry is not widely reported in the literature due to proprietary reasons. Tarmy and Coulaloglou (1992) discussed fundamental development and scale-up issues, illustrating these through the example of the coal liquefaction process. The traditional approach to scale-up is depicted in Figure 8.4, where kinetic constants obtained from laboratory data help correlate small pilot plant results. These pilot plant studies are then followed by demonstrations in larger units. They presented an evolving interactive scale-up strategy (Figure 8.5) that optimizes the use of resources, time, and technology.

8.9 Case studies on Scale-up projects of MB flotation from lab to industrial scale

8.9.1 Enhancing Coal Flotation Efficiency with Cyclonic MBs Flotation Column

8.9.1.1 Background and Objectives

In the coal mining industry, the efficient separation of fine coal particles is crucial for maximizing clean coal yield and reducing ash content. Traditional flotation cells often struggle with achieving optimal separation efficiency, particularly for fine coal particles under 0.5 mm in size. To address this challenge, researchers have explored the application of MB flotation columns in cyclonic devices (FCMCs) as an alternative technology for coal flotation. In a study reported by Lang (2017), they aimed to investigate the feasibility and effectiveness of scaling up the FCMC technology from laboratory-scale testing to industrial application. Specifically, researchers sought to evaluate the performance of FCMC in separating fine coal particles collected from the Xiezhuang coal preparation plant (CPP).

8.9.1.2 Methodology

Laboratory-scale experiments were conducted to compare the performance of FCMC with standard self-aeration flotation cells. Parameters such as ash content, clean coal yield, and flotation efficiency index were measured and compared between the two systems. Following successful laboratory trials, industrial-scale testing was undertaken using the FCMC-5000, with a column diameter of 5000 mm at the Xiezhuang CPP.

8.9.1.3 Results

(a) **Laboratory Testing:** Results from laboratory experiments demonstrated that FCMC outperformed traditional flotation cells in terms of majorly three parameters, such as clean coal yield, flotation efficiency index, and ash content reduction. The ash content of clean coal obtained by the bubble separation system was down by 1.34% points compared to flotation cells. Additionally, clean coal yield increased by 2.90%, and flotation efficiency increased by 6.64%.

(b) **Industrial-Scale Application:** Industrial-scale testing confirmed the superior performance of FCMC in real-world conditions. The FCMC-5000 achieved a further reduction of 2.70% in clean coal ash content. Clean coal yield and flotation efficiency index also saw substantial improvements, increasing by 15.10% and 18.99%, respectively.

(c) **Economic Impact:** The adoption of FCMC technology resulted in significant economic benefits for the Xiezhuang CPP. Annual profits increased by 46.773 million RMB following the replacement of traditional flotation cells with FCMC-5000. These findings underscored the potential of FCMC technology to enhance coal flotation efficiency and generate substantial profits for mining enterprises.

8.9.1.4 Economic Report

The economic analysis of implementing FCMC technology at the Xiezhuang CPP reveals substantial financial benefits and improved operational efficiency. By replacing standard self-aeration flotation cells with FCMC-5000, the CPP witnessed significant reductions in clean coal ash content and increased clean coal yield, leading to a substantial increase in annual profits. This report provides a comprehensive overview of the economic implications of adopting FCMC technology at the Xiezhuang CPP.

8.9.1.5 Cost–Benefit Analysis

(a) **Capital Investment:** The initial investment in acquiring and installing FCMC-5000 equipment represents the primary capital expenditure. This cost includes equipment procurement, installation, and associated infrastructure upgrades.
(b) **Operational Costs:** Operating expenses encompass routine maintenance, energy consumption, labor costs, and chemical reagents required for coal flotation processes. Additionally, ongoing training and technical support may incur supplementary expenses.
(c) **Revenue Generation:** The adoption of FCMC technology leads to increased clean coal yield and reduced ash content, resulting in higher-quality coal products. Higher-quality coal commands premium prices in the market, thereby increasing revenue generation for the Xiezhuang CPP.

8.9.1.6 Financial Impact

(a) **Cost Savings:** By reducing clean coal ash content and improving clean coal yield, FCMC technology mitigates losses associated with inferior coal quality. Moreover, operational efficiencies achieved through FCMC implementation may lead to cost savings in the long run.
(b) **Revenue Enhancement:** The production of higher-quality coal products enhances market competitiveness and allows the Xiezhuang CPP to capture premium prices. Increased revenue from coal sales contributes directly to overall profitability.

(c) Return on Investment (ROI):
Short-Term ROI: The initial capital investment in FCMC technology is offset by immediate improvements in clean coal quality and operational efficiency. Short-term ROI analysis demonstrates rapid recoupment of initial investment through increased revenue and cost savings.
Long-Term ROI: Over the long term, FCMC technology continues to generate sustained returns through ongoing improvements in clean coal yield, ash content reduction, and operational efficiencies. Long-term ROI calculations account for the cumulative impact of increased profitability over the equipment's lifespan.

8.9.1.7 Conclusions

The implementation of cyclonic MBs flotation column technology at the Xiezhuang CPP presents a compelling business case, with significant potential for enhancing profitability and operational efficiency. By reducing clean coal ash content, increasing clean coal yield, and driving revenue growth through higher-quality coal products, FCMC technology offers a pathway to sustainable economic success for CPPs. The substantial increase in annual profits underscores the transformative impact of FCMC technology on coal flotation processes and the broader coal mining industry.

8.9.2 Scale-up of MB-Based Pharmaceutical Wastewater Plant

8.9.2.1 Background and Objectives

The case study of Ponce-Robles et al. (2023) explores the feasibility and potential benefits of integrating O_3/MB systems as tertiary treatment in conventional WWTPs for agronomic purposes. The study aims to optimize system parameters to achieve pharmaceutical removal and *E. coli* reduction, enhance effluent quality, and assess energy consumption and associated costs.

8.9.2.2 Methodology

(a) **System Workflow and Ozone Production:** Through optimization, a flow rate of 2000 L/h and an ozone production rate of 15.9 gO_3/h yielded the best results. The presence of MNBs enhanced system efficiency by facilitating both direct O_3 oxidation and indirect oxidation that involve free $^\bullet OH$ oxidation mechanisms.
(b) **Pharmaceutical Removal Efficiency:** Removal percentages varied based on pharmaceutical characteristics. Cationic drugs, dipolar ionic compounds, and neutral compounds exhibited 85.68% removal efficiencies, while anionic compounds showed removal efficiencies between

57.62% and 77.34%. Compounds containing chloro and fluoro groups demonstrated lower removal efficiencies of 22.50%.

(c) **Effluent Quality and Storage:** Effluents met European regulation standards for wastewater reuse, demonstrating high agronomic quality. Quality improvements were observed during storage, with increased disinfection after 24–72 hours due to the persistence of MNBs. Proper storage could potentially reduce ozone and energy requirements.

8.9.2.3 Challenges and Future Considerations

Generation of Transformation Products (TPs): Further research is needed to assess the risks associated with TPs generated in O_3/MNBs processes, as some TPs may be more toxic than the original compounds.

Impact on Agronomic Level: The impacts of O_3/MNBs systems on plant growth, fertilizer effects, etc., require further examination to fully understand their implications at the agronomic level.

8.9.2.4 Conclusions

The implementation of O_3/MNBs systems in WWTPs for agronomic purposes offers promising results in pharmaceutical removal, *E. coli* reduction, and effluent quality enhancement. Optimization of system parameters, understanding pharmaceutical characteristics, and addressing energy consumption challenges are crucial for maximizing benefits and ensuring sustainable implementation. Further research is warranted to assess risks associated with TPs and understand the broader impacts on agronomic practices.

Nomenclature

Notation		SI unit
c_f	final concentration of pollutant	mol/L
c_i	initial concentration of pollutant	mol/L
D_i	impeller/agitator diameter	m
D_P	diameter of pipe	m
e_v	loss due to fittings (friction)	m
E_{EO}	electric energy requirement per order	kWh/m^3/order
E_M	electric energy per mass	kWh/kg
f	friction factor	unit less
f_1	function obtained graphically	unit less
Δh	height through which the fluid is lifted	cm

Notation		SI unit
H_p	power	HP
k_0	rate of consumption	concentration$^{1\text{-order}}$/time
k_La	mass transfer coefficient	1/s
k_La_{20}	dissolution rate of oxygen in water	1/s
L_P	length of pipe	cm
M	molar mass of pollutant	g/mol
N	rotational speed of the impeller/agitator	rpm
N_P	power number	unit less
P_0	power supply for un-gassed system	W
P_F	power to work on the fluid	W
P_G	power supply for gassed system	W
P_o	power supplied to the impeller/agitator	W
P_{MBG}	power supply for MB generator	W
P_{motor}	power supply for motor	W
P_{pump}	power supply for pump	W
$P_{recycle\ pump}$	power supply for recycle pump	W
Q	volumetric gas flow rate	m^3/s
Q_{H_2O}	flowrate of water	m^3/s
Q_{air}	flowrate of air	m^3/s
v	fluid average velocity	m/s
V_{MBG}	volume of MB generator	m^3
W	pump shaft work	W

Greek Nomenclature

Notation		SI unit
βi	dimensionless parameter	
ε_D	energy dissipation rate	m^2/s^3 or W/kg
μ_L	liquid viscosity	$N.s/m^2$
ρ_L	liquid density	kg/m^3

References

Bolton, R.J., Bircher, G.K., Tumas, W., and Tolman, A.C. (2001). Figures-of-merit for the technical development and application of advanced oxidation technologies for both electric and solar-driven systems. *Pure and Appl. Chem.* 73: 627–637.

Degaleesan S., Fluid dynamic measurements and modeling of liquid mixing in bubble columns. D.Sc. thesis, Washington University, St. Louis, MO, 1997.

Harun, M.H.C. and Zimmerman, W.B. (2019). Membrane defouling using microbubbles generated by fluidic oscillation. *Water Sci technol. Water supply.* 19: 97–106. ISSN 1606-9749.

Hensirisak, P., Parasukulsatid, P., Agblevor, F.A. et al. (2002). Scale-up of microbubble dispersion generator for aerobic fermentation. *Appl. Biochem. Biotech.* 101: 211–227.

Inga, J.R. and Morsi, B.I. (1997). Effect of catalyst loading on gas/liquid mass transfer in a slurry reactor: a statistical experimental approach. *Can. J. Chem. Eng.* 75: 872–881.

Jiang, L., Wang, H.L., Liu, W.Y. et al. (2022). HiGee microbubble generator: (i) mathematical modeling and experimental verification of the energy dissipation rate. *Ind. Eng. Chem. Res.* 61: 16823–16831.

Khuntia, S., Sinha, M.K., and Singh, P. (2021). Theoretical and experimental investigation of the mechanism of the catalytic ozonation process by using a manganese-based catalyst. *Environ. Technol.* 42: 632–639.

Lang, X. (2017). Application of cyclonic microbubble flotation column in Xiezhuang coal preparation plant: from laboratory to industrial scale. *IJCP&U* 40: 403–417.

Ochoa, E.D.O., García, M.C., Padilla, N.D., and Remolina, A.M. (2022). Design and experimental evaluation of a Venturi and Venturi-vortex microbubble aeration system. *Heliyon* 8: e10824.

Parmar, R. and Majumder, S.K. (2016). Mineral beneficiation by ionic microbubble in continuous plant prototype: efficiency and its analysis by kinetic model. *Chem. Eng. Sci.* 142: 42–54.

Perry, R.H., Green, D.W., and Maloney, J.O. (1984). Perry's Chemical Engineers Handbook, vol. 5, 6–16. New York: McGraw-Hill.

Ponce-Robles, L., Pagán-Muñoz, A., Lara-Guillén, A.J. et al. (2023). Full-scale O_3/micro-nano bubbles system based advanced oxidation as alternative tertiary treatment in WWTP effluents. *Catalysts* 13: 188.

Singh, A., Dalvi, S., Sinha, M.K., and Khuntia, S. (2024). Magnetite nanoparticle mediated ozonation of ciprofloxacin wastewater: energy calculation and optimization of parameters. *Chem. Eng. Commun.* https://doi.org/10.1080/00986445.2024.2362815.

Su, S., Huang, F.Y., Liu, B.B. et al. (2022). A feasible strategy for deeply separating low concentrations of molybdenum from tungstate solutions with a high-efficiency microbubble floating-extraction concept. *ACS Sustain. Chem. Eng.* 10: 146–158.

Tarmy, B. and Coulaloglou, C. (1992). Alpha-omega and beyond industrial view of gas/liquid/solids reactor development. *Chem. Eng. Sci.* 47: 3231.

Wilkinson, P.M., Spek, A.P., and van Dierendonck, L.L. (1992). Design parameters estimation for scale-up of high pressure bubble columns. *AIChE J.* 38: 544.

Xia, Z. and Hu, L. (2019). Treatment of organics contaminated wastewater by ozone micro-nano-bubbles. *Water* 11: 55.

Zimmerman, W.B. and Kokoo, R. (2018). Esterification for biodiesel production with a phantom catalyst: bubble mediated reactive distillation. *Appl. Energy* 221: 28–40.

Index

f

g

h

n

Printed and bound by CPI Group (UK) Ltd, Croydon, CR0 4YY
17/09/2026
14964330-0001